Ground subsidence

**INSTITUTION OF CIVIL ENGINEERS
LONDON 1977**

Produced and distributed by Thomas Telford Limited for the Institution of
Civil Engineers, PO Box 101, 26–34 Old Street, London EC1P 1JH

ISBN: 0 7277 0046 4

Text set in 11/12 pt Photon Times, printed by photolithography and bound
in Great Britain at The Pitman Press, Bath

Contents

Foreword

In 1974, the Publications and Meetings Committee of the Institution of Civil Engineers set up a small working party which advised that the current *Report on mining subsidence*, written between 1952 and 1959, was out of date and excluded problems of subsidence caused by natural cavities and disused workings.

Considerable advances have been made during the intervening period in locating areas of potential subsidence, in predicting the effects of subsidence and in foreseeing the tolerance of different types of civil engineering works and structures to ground subsidence.

The decision was therefore taken to prepare a new *Report on ground subsidence.* In the interests of expedition it was decided to appoint a small editorial board to organize the new report, preparing a structure with named authors to prepare chapters or part chapters. In the event substantial portions of the report have been largely written by members of the Editorial Board, who were: A. M. Muir Wood (Chairman), Sir William Halcrow & Partners; I. H. McFarlane, Mitchell, McFarlane & Partners; Lt Col. F. R. Oliver; G. K. Raven; Gilbert S. Senior, Consulting Engineer, S. Thorburn, Thorburn & Partners; P. B. E. Thompson (Secretary).

The Council of the Institution is greatly indebted to the following who have also contributed to the report and reviewed draft chapters: G. Archibald (National Coal Board); J. E. Cheney (Buildings Research Establishment); J. B. A. Day (Lancashire County Council); M. H. de Freitas (Department of Geology, Royal School of Mines, Imperial College of Science and Technology). E. M. Gosschalk (Sir William Halcrow & Partners); C. J. F. Jones (West Yorkshire Metropolitan County Council); F. W. Newborn (Deputy Estates Manager, National Coal Board, South Yorks Area); R. J. Orchard (National Coal Board); J. F. S. Pryke (J. F. S. Pryke & Partners); M. C. Purbrick (Chief Civil Engineer, British Railways Board); R. Rukin (River Drainage Officer, Yorkshire Water Authority, Rivers Division, Land Drainage, Leeds); A. Smith (Mining Engineer, Department of Planning, Engineering and Transportation, Wakefield); M. Staikes (Mining Department, British Waterways Board); G. Walsh (West Yorkshire Metropolitan County Council); P. F. Winfield (George Wimpey & Co. Ltd).

The report sets out to provide guidance to good practice for the civil engineer who is not a specialist in matters of ground subsidence; it also provides an introduction to the extensive literature on more specialized aspects of ground subsidence.

Chapter 1
Causes of ground subsidence

NATURAL CAVITIES
Susceptible geological conditions

Cavities in sedimentary deposits may occur either from the open texture of the material or from internal differential solution subsequent to deposition. Rock soluble in circulating groundwater is most susceptible to the formation of large cavities. Here it is necessary to distinguish between the two most widespread soluble constituents of rock: chlorides and carbonates.

Chlorides occur mainly as evaporites, which are predominantly rock salt (NaCl). Evaporites behave plastically at depth; consequently, solution will not lead to formation of cavities except where erosion is local and where the salt is overlain by relatively competent rock.

Studies have been made of the rate of erosion of carbonates by circulating groundwater. Lower temperatures and consequent higher carbon dioxide concentrations result in more corrosive water, and a greater rate of attack by water flowing through joints and fissures. Subsequent to such attack, percolation of seawater or other brine containing magnesium salts may cause dolomitization of the limestone, the process starting adjacent to the flow. Dolomitization entails the substitution of dolomite (the double carbonate of calcium and magnesium) for calcite (calcium carbonate). Where replacement is selective, subsequent solution may lead to a strong rock skeleton of a highly cavernous (karstic) rock leading to the caves and rock pinnacles of karst country.

Unconsolidated sediments (the engineer's soil) may have been subjected, before consolidation, to mineral cementation. If such sediments, subsequent to such cementation, have been free of circulating water, their inundation or partial wetting may lead to collapse of the soil structure. These conditions are most likely to be found with wind-blown sediments and of these loess is the most widespread example of a collapsing soil. A somewhat similar

phenomenon, that of base exchange, may give rise to sensitive clays, particularly where marine deposits have been subjected to leaching of the
sodium ion. Here, however, the effect is that of dramatic loss of strength on
disturbance or overloading.

Intrusive igneous rocks (i.e. those which cooled relatively slowly under a
cover of pre-existing rock) are massive, crystalline materials such as granite
and dolerite, whose cavities will be confined to intercrystalline microfissures.
Weathering may enlarge these fissures, mainly by solution, giving rise to
porous residual soils (Wallace, 1973). Large-scale discontinuities may occur
as joints in sets that are ubiquitous and systematic, e.g. columnar jointing in
sills and dykes. Weathering and erosion are most active here and can widen
joints by many centimetres.

Extrusive rocks are fine-grained on account of rapid cooling and include
lavas. Apart from microfissures, entrapped gas may give rise to voids which
may be filled with weak hydrous minerals such as zeolites, as in amygdaloidal
lavas (almond shaped cavities), or may remain empty, as in pumice. Many
extrusive rocks weather rapidly to highly porous residues which are susceptible
to collapse under load. Large-scale cavities in unweathered lava flows are mainly associated joints which are columnar, as with intrusive rocks. The largest
type of cavity, but not the most common, is the lava tunnel; the crown of such a
tunnel may be weak if composed of thin lava flows or of weathered material.

The recrystallization that occurs in rocks altered by combined heat and
pressure reduces porosity so that unweathered metamorphic rocks present
no natural form of cavity of engineering significance. The weathering of
joints is comparable to that of igneous rocks but joint systems are less
regular for metamorphism that occurs at low pressure.

The existence of cavities in rock is favoured by tectonic faulting and
folding. Where a fault is marked by a fault zone rather than a thin shear
plane, the filling material may be angular, cavernous and resemble a porous
breccia. Faults may be subsequently healed by crystallization of the host
rock or by mineralization, whereby the breccia cavities are filled and
cemented.

'Piping' is a phenomenon familiar to the civil engineer, in relation to
engineering works, whereby the pressure gradient of a percolating fluid is
sufficient to carry particles into suspension. Piping may also occur naturally, where high hydraulic gradients have been associated with erodible
material overlain by more competent and relatively impermeable material or
by ice. Subsequent geological history may lead to filling of the pipe and the
most likely situation of residual cavities is alongside valleys deepened by
glacial action.

Conditions of collapse

The age of most natural cavities is such that collapse is likely to occur only where there is an adverse change of factors affecting natural or artificial stability (Lajtai and Lajtai, 1975). The commonest factors are those of changes in groundwater and seismic activity, each leading to variation in effective stress in the ground.

Lowering or raising or piezometric pressure will respectively cause increase in effective pressure or increase in shear stress in relation to shear strength, either change being capable in particular circumstances of causing collapse of the natural arch over a cavity. Rapid increase in groundwater flow or sudden change in the chemistry of groundwater may lead to reinvigorated solution of the type previously discussed, leading ultimately to collapse where the rock becomes overstressed. Release of oil or gas under pressure may lead to similar results (Poland, 1971).

The effects of seismic shock are twofold: first, the stress changes caused by ground acceleration; second, the changes in pore-pressures (air, oil or gas) in the ground, which will be accentuated by consequential collapse of the ground.

A spectacular example of subsidence is seen where lowering of the groundwater by mining in the Western Transvaal has caused collapse of large caverns in the dolomite where this is relatively close to the surface, resulting in subsidence damage over areas up to 500 m in diameter. In South-east England a similar phenomenon on a smaller scale is caused by periodical collapses of unconsolidated sediments overlying vertical pipes in the chalk, caused by fluctuation in groundwater or in surface loading (West and Dumbleton, 1972).

BASIC CAUSES OF MAN-MADE UNDERGROUND CAVITIES

Principles

The principles of the extraction of minerals other than coal have the same effects as those which operate within coal mining. When any minerals are excavated beneath the Earth's crust, the superincumbent rock may fill the cavities where the minerals have been excavated, thus leaving a depression on the surface. If any structures have been or are to be erected over such workings, then movement could occur which could cause damage.

Many minerals have been extracted by various methods: these minerals include coal, lead, tin, various types of metalliferous mineral, gypsum, anhydrite, salt, stone, chalk and/or flints. The most frequent, however, is the extraction of coal, gypsum, anhydrite and salt. Past operations have taken place in various parts of England by the working of lead veins and of tin, particularly in Cornwall, but such workings are practically extinct. The broad outline of the effects of such mineral extraction has been mentioned earlier.

The working of salt, either by underground extraction methods or by pumping, lowers the surface in a manner different from that resulting from the extraction of coal, and those precautions required in the design and construction of buildings in salt mining areas give rise to much more detailed problems.

Gypsum and anhydrite are extensively worked in Sussex, Nottinghamshire and Yorkshire, but the method of working is so designed that little or no movement is recorded on the surface: large pillars are left to provide support for the surface. These two mineral deposits occur in very much greater thicknesses than coal does and consequently economic working does not necessarily limit support to small pillars. Chalk and/or flints have been mined by methods similar to those used for the working of the minerals mentioned in the foregoing, by the application of pillar and stall methods. There is evidence of this in Kent, known as the Chislehurst Caves, and the Chalk Mines in areas in and around Bury St Edmunds, Suffolk.

History and effects

The story of the development of mining commerce in coal in England and Wales probably begins with a flint axe which was found stuck into a bed of coal in an old coal mine in Monmouthshire. This has been regarded by some as incontestable evidence of coal mining having been practiced by the Ancient Britons; and ornaments made of cannel have been found in glacial drift, but this is very slender evidence on which to base the supposition of the working of coal by the tribes of Britain. In Anglo Saxon times, the supposed word for coal *(graefa)* is mentioned in the Saxon corner of the Abbey of Peterborough, AD 852, but there is much doubt as to the meaning of the word *graefa*. It may mean larger coal; it might even mean gravel. The outstanding fact remains that no reference is made to coal in the Domesday Book, that great survey carried out about 1085 by the order of William the Conqueror. As the survey is such a minute and accurate record, which refers to lead mining works and iron mines, as well as particulars of the salt industry, reference would surely have been made to coal mining, if carried on, however small the scale. Probably the first step in getting coal was by the simple digging of coal exposed at the surface and it would then come to be sold as an article of commerce.

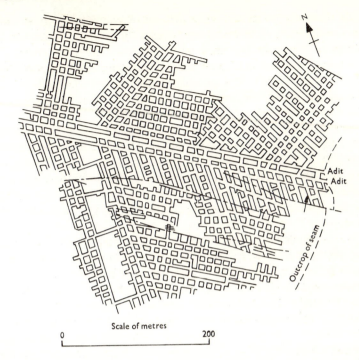

Fig. 1. Typical example of pillar and stall working in coal

In AD 1215, an act was passed which safeguarded the rights of surface owners, for in 1259, 44 years after the signing of the Magna Charta, the Freemen of Newcastle-on-Tyne were granted a charter for the liberty to dig coal, and coal was definitely worked in Warwickshire in 1275. In 1306, it was regarded as a very dangerous innovation, and Parliament was petitioned to prohibit the burning of sea coal in London and the suburbs to avoid 'sulphorous smoke and saviour of the firing' and the destruction of furnaces and kilns was authorized to prohibit its use. The first step in mining proper was simply to take coal from the seam where exposed at the surface through a few metres of overlying strata. The term 'bell pit' was derived from the fact that the excavations were widened out where they struck the seam, so as to allow as much coal as possible to be removed without causing the upper strata to fall in.

It is recorded that in the 14th century, coal was mined by shafts and horizontal adits (a tunnel driven in by the hillside at such a level as to drain off water from the workings). Such methods were adopted in shallow seams of coal and pillars were left in to support the surface ('pillar and stall workings'). The earliest method of mining coal in Great Britain was probably employed by the Romans, but it was not until the days of the Industrial Revolution that any extensive workings were undertaken. It has been suggested that small areas of coal were mined in Northumberland in the year 1110 but there are no details available.

As might be expected, the earliest miners looked to the shallower seams for their output, and pillar and stall mining, a method of partial extraction, was evolved. A system of headings or roadways were driven in the seam, usually at right angles to each other (Fig. 1). The width of headings was such that the intermediate roof would not readily collapse and the pillars would be of sufficient dimensions to support the upper strata and overlying surface.

The ultimate intention, after reaching the boundary of the mine, would be to retreat to the shaft extracting the pillars and allowing the roof to fall into the goaf or open area where the seam had been extracted. Unfortunately, this was not always done, and it is these shallow partial extraction workings which are usually uncharted and give rise to anxiety when their presence is suspected. The deterioration of roof strata may cause a collapse of the surface, but only after a great number of years. The subsidence engineer can do little to determine their effect at the surface and on surface structures, even if a drilling programme proves the extent of the workings and the associated pillars, as only a close examination of these would disclose their condition, and thus their load bearing capabilities.

As mining techniques advanced, the 'long wall' system was developed, and this is illustrated in Fig. 2. More or less complete extraction is achieved as working fronts or faces, varying between fifty and three hundred metres in width are advanced by taking cuts at regular intervals. These cuts are done by machines which vary from $1 \cdot 3$ to $1 \cdot 8$ m in width. Ideally, the faces are planned in sequence from a main development face which explores and opens up the available reserves in the seam.

A modern development of partial extraction, known as strip mining, is based on the long wall system, each long wall face, having a width carefully related to the depth of the seam from the surface, is worked alternately, leaving long pillars of coal of the same adequate width between faces for support of the surface. This method results in a shallow and even subsidence basin accompanied by very moderate strains and results in little or no damage to surface structures. A technique has recently been evolved whereby two roads, approximately 200 metres apart, are driven for a given distance and

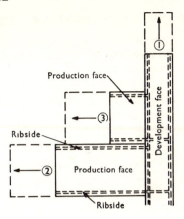

Fig. 2. System of long-wall working

then the coal extracted in a reverse manner. This system is known as retreat mining.

PROBLEMS ASSOCIATED WITH MINING AND GROUNDWATER

Flooding

One of the serious problems which is associated with mining, where old workings have taken place, is the fact that water may be present in such old workings, and very stringent precautions are thus employed by drilling ahead of active workings to prove whether any water is present or not. It should be appreciated that where old workings have been full of water, the surface may be afforded hydrostatic support, and if such water is taken away, then there is a possibility that the goaf resulting from the previous extraction of the coal will cause lowering effects of the superincumbent rocks, and thus cause damage at the surface.

REFERENCES

Lajtai, E. Z. & Lajtai, V. H. (1975). The collapse of cavities. *Int. Jnl Rock Mech. Mining Sci. Geomech. Abs.* **12,** 81–86.

Poland, J. (1971). Subsidence and its control in: Underground waste management and environmental implications. *Am. Assoc. Petroleum Geologists,* Memoir 18, 50–71.

West, G. & Dumbleton, M. J. (1972). Some observations on swallow holes and mines in the Chalk. *Q. Jnl Engng Geol.* **5**, 171–177.

Wallace, K. B. (1973). Structural behaviour of residual soils of the continually wet Highlands of Papua New Guinea. *Géotechnique* **23**, No. 2, 203–218. Discussion: *Géotechnique* **23**, No. 4, 601–603 and **24**, No. 1, 101–106.

Chapter 2
Identification and investigation

POTENTIAL PROBLEMS

The potential problems are those where uncharted workings have taken place and no plan records are in existence. This appertains to all forms of underground mining and is not associated merely with coal mining. The following paragraphs will give sufficient information to enable the engineer to seek and obtain information where it is known.

Information can be obtained to ascertain whether a potential problem may arise from natural and man-made cavities, by reference to the Geological Survey Maps which date from 1857, this being the first edition, and other editions from 1929 onwards. These maps are available for inspection for the Midlands and the South of England, and for Wales, at the Institute of Geological Sciences, London; for those in the North, from the Institute of Geological Sciences, Halton, near Leeds; and for Scotland, at the Institute of Geological Sciences, Edinburgh. It is also possible to examine the field notes from which these maps were prepared and all journals which are in the possession of these institutions. Valuable information can thus be obtained in this direction.

IDENTIFICATION OF THE PROBLEM

Another source of information is that provided by the mineral valuer when a local authority is obtaining a valuation of land. The mineral valuer for that particular district issues a report on the stability of such land, compiled from records in his possession, and while it is confidential, local authority representatives will be able to obtain such information as may be available by direct contact with such officers.

Records of extraction

The National Coal Board has collated all the old records of mine workings which are known to exist at their Area Offices, and these records may be inspected by the general public. It is possible to obtain copies of these working plans on the payment of a small fee.

Mining engineers in private practice have recently been asked to deposit all their old records with the National Coal Board so there is one central source where such information can be obtained.

With regard to minerals other than coal and oil shales, a *Catalogue of abandoned mines* covering all minerals except the above is available for inspection at the Mining Records Office, Health and Safety Executive, Department of Employment, London, and other information can also be obtained from HM Divisional Inspectors of Mines and Quarries, whose offices are situated in various parts of the United Kingdom. The records of the minerals other than coal, which are extremely well catalogued, are available for inspection by the general public free of cost. Copies of such old plans may be obtained on the payment of a small fee.

METHODS OF INVESTIGATION

There are many ways of investigating the presence of old mine workings and these may be classified as follows:

(*a*) visual inspection;
(*b*) aerial photography;
(*c*) drilling techniques;
(*d*) geophysical methods.

Visual inspection

Where shallow workings in any minerals have taken place, mounds and undulations usually occur and can be seen by a visual inspection of the area. In particular, bell pit workings, as described in Chapter 3, result in hollows which give a clear indication that some form of unnatural movement has occurred.

Aerial photography

The application of aerial photography usually pinpoints areas of depressions and other phenomena which would indicate that some unusual form of surface movement has occurred. Particularly in the planning of some of the motorways in the UK and abroad, the application of aerial photography has been of very great assistance.

Drilling techniques

Exploration is normally carried out in the very early stages as a method of preliminary investigation of the strata, both to locate man-made cavities and to determine the nature and characteristics of the ground. This objective can be achieved by the application of:

(*a*) shell and auger work;
(*b*) core drilling.

Core drilling is not universally used for site investigation methods, but it can be adapted for this work by using a drill which would produce core samples which can be tested in the same way as the shell and auger methods.

There are limitations, however, to the application of both of these as they are expensive and absorb a great deal of time in carrying out the work, and shell and auger methods cannot easily penetrate hard rock strata.

This is the most useful method of establishing whether shallow workings exist beneath a site, and it can be achieved by drilling small diameter boreholes, normally 75 mm in diameter, by using either water flush or air flush methods where no core of the strata is required. The drill blows out debris from the top of the hole and indicates, by examination of colour, texture and feel, what type of stratum is being drilled. It is essential that an engineer who is familiar with ground geology should supervise such work, as the interpretation of the debris from the hole is the key to the ultimate description of those strata.

If cavities or broken ground exist, the engineer, or driller, can then record, by the behaviour of the drill, the type and depth of cavitation. This technique has now become a well established method of investigation and various drilling companies have become expert in this field; it is a cheap and quick method of ascertaining the type of strata beneath the surface. The holes to be drilled are laid out on a given pattern and, therefore, sections can be drawn either longitudinally or horizontally across a site. By this method, the engineer can then determine what areas require grouting or alternative means of stabilization.

Geophysical methods
The application of this system is of recent origin and will indicate a variation in the type of stratum, being more particularly useful in determining the presence of major faulting systems. It has been found to be of great benefit, e.g. in tracing the major faults with throws of over 10 m in the new Selby Coalfield in North Yorkshire. Its application for locating uncharted mine workings of any description is not as reliable as the drilling techniques mentioned above, and a great deal more research is required in this field before it can be adopted for this purpose.

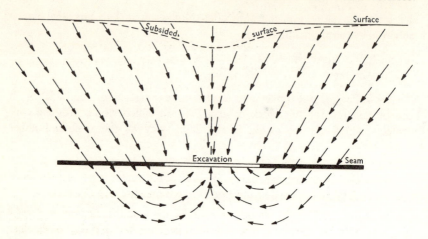

Fig. 3. Pseudo-plastic flow: characteristic lines of ground movement (after Grond)

MONITORING AND CONTROL OF GROUND MOVEMENT

The work of the subsidence engineer is concerned with the quantitative estimate of predicted ground movement arising from a given set of circumstances. In doing this, he will base his estimates on the results of
numerous precise levellings and measurements made on the surface over
workings, not only in his own coalfield, but throughout the country. The ultimate aim must, of course, be to determine the effect of the movements
predicted on any given type and size of structure, and it is this information
which the civil or structural engineer requires.

There has been a great effort by the mining engineer to develop
techniques within the mines to control the movement of strata and thus limit
the amount of subsidence which takes place on the surface. This has been
achieved by close liaison with the subsidence engineer by limiting the size of
workings in relation to the superincumbent rocks and the area over which
limited ground movement could be permitted.

THE NATURE OF GROUND MOVEMENT AND
SUBSIDENCE EFFECTS

It has been proved by observations by Professor Grond of the Netherlands,
and more latterly by Wardell and Orchard, that when a cavity is made underground, the surrounding strata flow inwards in all directions in a pseudo-
plastic manner, as shown by the general lines of movement in Fig. 3, and it
will be appreciated that movement within the trough is three-dimensional. It

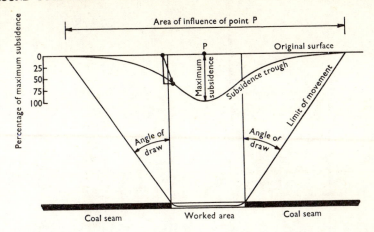

Fig. 4. Area of influence of point P. In this case the width/depth ratio is 0·4 so that the half subsidence point is over the edge of the working face

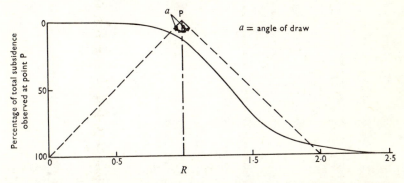

Fig. 5. Advance of working across area of influence of point P in terms of radius of influence

may be seen that the resulting surface deformation which takes the form of a shallow trough, covers an area considerably larger than the excavated area. Fig. 4. is derived from Fig. 3 and illustrates the angle of draw which may be defined as being the angle between the line from the edge of the excavated area, normal to the seam, and the line joining the edge of the excavated area to the point of zero effect at the surface. It is not easy to define this latter point because of the asymptotic nature of the subsidence curve and the more precise the observations, the greater the angle of draw will appear to be. It has been found from experience and research in practice carried out that this figure varies between 25° and 35°.

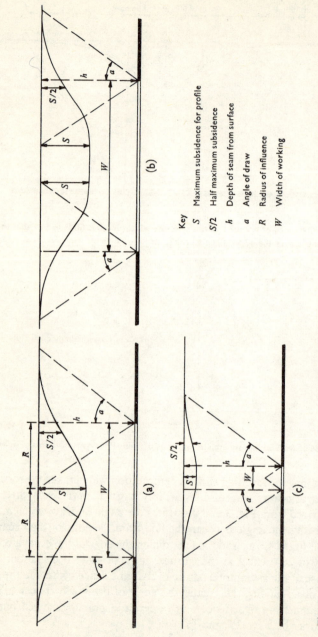

Key

S	Maximum subsidence for profile
$S/2$	Half maximum subsidence
h	Depth of seam from surface
a	Angle of draw
R	Radius of influence
W	Width of working

Fig. 6. Development of subsidence curve according to the width/depth ratio: (a) width/depth ratio = 1·4; (b) width/depth ratio > 1·4; (c) width/depth ratio < 1·4

Up-to-date prediction

If one takes the surface point P and using the angle of draw projects an area on to a given seam of coal, the result is the 'area of influence' or critical area of the point P (Fig. 5), and in the case of a level seam, this will normally be a circle. The following important points are now apparent:

(a) Arising out of the definition of the angle of draw, no coal worked outside the area of influence can have any effect on point P.

(b) Unless all the coal within the area of influence is extracted, the point P will not undergo full subsidence. This principle will, of course, apply when one substitutes a structure for point P. Fig. 6 shows the effects on point P of the workings of varying width. Fig. 6(a) shows a face having a width equal to the diameter of the area of influence. This will be $2\,h\tan a$ where h is the depth and a the angle of draw, so that if one assumes an angle of draw to be 35°, the width/depth ratio is 1·4. Maximum subsidence of the point P will occur and the subsidence trough will not extend beyond the limit shown. Fig. 6(b) illustrates the effect of a face having a width greater than the diameter of the area of influence so that the width/depth ratio is greater than 1·4. P has not subsided further but the subsidence trough has widened and developed, giving a flat-bottomed profile.

(c) The effect of a face narrower than the area of influence is shown in Fig. 6(c). The width/depth ratio is less than 1·4 and unworked coal remains within the area of influence, P has not fully subsided, resulting in a shallow profile with subsidence over the ribside only a little less than that over the centre of the face.

Other things being equal, vertical lowering at the surface is in proportion to the thickness of the underground excavation, and it is important to be able to determine the percentage which will be transmitted to the surface as maximum subsidence at the centre of the trough. The National Coal Board Subsidence Engineer analysed a very large number of observed cases from all parts of the coalfields in Great Britain, and deduced a partial subsidence curve from which may be read the relevant subsidence factor for any width/depth ratio up to 1·4, after which maximum subsidence does not increase. Knowing the width of the workings, the depth of the seam, the thickness of the seam, and having estimated the angle of draw, the subsidence engineer can then plot the limit of movement, together with the position and the amount of maximum subsidence, but in order to complete the subsidence profile, more plotting points are needed.

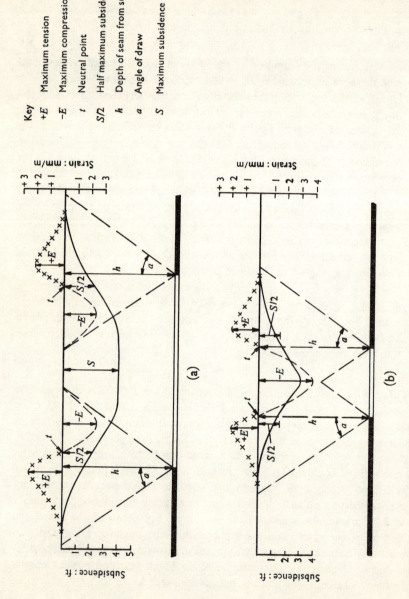

Fig. 7. Strain profiles: (a) width/depth > 1·4; (b) width/depth < 1·4

Key
+E Maximum tension
−E Maximum compression
t   Neutral point
S/2 Half maximum subsidence
h   Depth of seam from surface
a   Angle of draw
S   Maximum subsidence

Further research, again based on a larger number of observed cases, resulted in a graph which shows that the half subsidence point migrates from outside the ribside when the width/depth ratio is less than 0·4 to a fixed proportion of the depth over the extracted area which is reached when the width/depth ratio is equal to 1·4.

Referring to Figs 3 and 4, it will be appreciated that, as the subsidence trough develops, the centre will subside vertically, while the remainder will move inwards towards the centre of the workings, resulting in both vertical and horizontal movements. The differential horizontal movements give rise to straining of the overlying strata, the inward movements towards the fixed centre resulting in a zone of compression at the centre, whilst the movements away from the fixed extremities of the trough result in two zones of tension.

It is the strains resulting from subsidence, rather than the vertical tilts which cause the majority of damage to structures but, nevertheless, tilts in bridge structures are of major importance. It is highly important for the subsidence engineer to be able to quantify the anticipated strains and to plot the position of the zones of tension and compression. Simple formulae for the determination of maximum strain figures have been derived from observed data by authorities on this subject, based on the fact that strains are proportional to the thickness of the extraction and inversely proportional to the depth of the workings, and it has been proved that the transition point from tension to compression coincides for practical purposes with the half subsidence point. Orchard's research has shown that the points of maximum strains bear a definite relationship to the ribside, depending on the width/depth ratio, and he has arrived at the position where, knowing this latter factor, the position of maximum strain points may be read from graphs.

Fig. 7 shows the strain profile superimposed on the diagram used to illustrate the subsidence profile. Fig. 7(a) illustrates the position where the width/depth is greater than 1·4; it will be noticed that on either side of the subsidence trough, is a tension zone accompanied by a compression zone of almost equal intensity, while the centre of the trough is strain free. Fig. 7(b) shows the position where the width/depth ratio is 0·7 or less and the important point here is that not only is there no strain free zone over the centre of the trough, but the intensity of the compressional strains is greater than that of tensional strains.

The National Coal Board, through research carried out by Orchard and others, has produced a subsidence engineers' handbook (National Coal Board, 1975), which explains in detail all these points, together with symbols and definitions which have been derived, and by the application of the data in this handbook, predictions of movement at the surface can be reasonably accurately calculated.

Efforts have been made to solid-stow the goaf by pneumatic methods of filling this space with compressible material to reduce the ultimate movement at the surface. However, it is rarely possible to reduce this amount which can range up to 90% of the thickness of the seam, by as much as 50%, and this operation is difficult and extremely expensive. It is not used in general practice.

A paper on computer prediction of ground movements due to mining subsidence was presented to the Institution of Civil Engineers by Jones and Bellamy (1973). By the application of these techniques, it is possible, by having available certain mining data, to estimate the movement of an individual point in areas unaffected by faulting systems.

Significance of geological structure

The effects of ground movement are determined by the nature of the overlying strata, faulting systems and fractures, and breaks within the strata. In coal mining generally, the extraction of seams of coal, particularly those at deeper levels, will cause the overlying rocks to fracture and break and possibly become self-supporting for a number of years without any significant effects upon the surface. Breaks in the strata by a further disturbance may then cause large areas of rock deformation, with the ultimate result of many areas being subject to stress and strains and ultimate damage. Faulting, which is caused by a break in the earth's crust, has serious effects where such displacement comes to the surface.

Variations in the types of stratum lying between the seam and the surface do not have any effect on the maximum amount of subsidence. Sand or alluvium may cause the movements to spread out further, and certain types of rock which are present, such as magnesian limestone, tend to fracture badly and can have a profound effect on the intensity of differential displacements, thus causing serious damage at the surface. Breaks in the ground due to previous working of other seams can have an influence where these other seams have been only partially extracted, thus causing differential settlement.

REFERENCES

Berry, D. S. (1977). Progress in the analysis of ground movements due to mining. *Proc. Conf. Large Ground Movements and Structures, Cardiff.*

Grond, G. J. A. (1950). Disturbances of coal measures strata due to mining activities. *Iron Coal Tr. Ref.* **160,** 1323, 1377, 1445; **161,** 37, 85, 135, 197, 249, 295, 353, 394. (Also in *National Coal Board Production Dept. Info. Bull.* 51/49 and MG 51/2 (1951).

Grond, J. H. A. (1953). *A critical analysis of early and modern theories of mining subsidence and ground control*. Department of Mining, Leeds University.

Grond, G. J. A. (1957). Ground movements due to mining. *Colliery Engng* **34,** 157–197.

Jones, C. J. F. P. & Bellamy, J. B. (1973). Computer predictions of ground movements due to mining subsidence. *Géotechnique* **23,** No. 4, 515–530.

National Coal Board (1975). *Subsidence engineers' handbook*. Mining Department, National Coal Board, Hobart House, London.

Orchard, R. J. (1953). Recent developments in predicting the amplitude of mining subsidence. *Jnl. Roy. Instn Chart. Surv.* **33,** 864.

Orchard, R. J. (1957). Prediction of the magnitude of surface movements. *Colliery Engng* **34,** 455.

Orchard, R. J. (1964). Partial extraction and subsidence. *Min. Eng.* No. 43, April.

Orchard, R. J. (1964). Surface subsidence resulting from alternative treatments of colliery goaf. *Colliery Engng*, October.

Shadbolt, C. H. (1977). Mining subsidence—historical review and state of the art. *Proc. Conf. Large Ground Movements and Structures, Cardiff.*

Piggott, R. J. & Eyron, P. (1977). Ground movements arising from the presence of shallow abandoned mine workings. *Proc. Conf. Large Ground Movements and Structures, Cardiff.*

Wardell, K. (1953). Some observations on the relationship between time and mining subsidence. *Trans. Instn Min. Engrs* **113,** 471. (Discussion, p. 799).

Wardell, K. (1957). The minimisation of surface damage. *Colliery Engng* **34,** 361.

Chapter 3
Preventive measures

GROUTING
Introduction

Grouting involves the filling of cavities in the strata, either naturally formed or resulting from the extraction of minerals by underground workings, by injection with a suitable inert material in a liquid suspension. The methods of grout injection referred to in this chapter relate to the grouting of cavities resulting from the extraction of coal and other minerals at relatively shallow depths from the surface. The injection methods used for the grouting of fissures and natural cavities (for example, foundations of dams and tunnel excavation) have been the subject of many papers over the past decade and will not be dealt with in this report. Grouting will be undertaken only where excavation is not a practicable proposition or where costs of alternative methods of stabilization, such as structural measures, are significantly more expensive.

Plant

Drilling plant. Grout holes are normally drilled by track mounted rotary percussive drilling rigs. These rigs can readily cope with adverse site conditions, such as soft ground and limited working space. The high drilling rate obtained from this type of rig is also an important factor and excellent results can be achieved using air flush.

If the initial exploratory drilling of the site reveals the presence of large cavities in the mine workings then the grout boreholes should be 100 mm diameter to enable pea gravel to be introduced as required during the grouting operations. Where the site investigation indicates that large areas of mine workings are collapsed or partially collapsed with extensive areas of broken ground in the immediate overlying strata, then the minimum diameter of the boreholes should be 70 mm. If the diameter is less then there is a high risk that blockages will occur in the boreholes when broken ground is encountered. Where surface deposits of sand, gravel and fill are present, temporary casing of the borehole may be required. Provision for temporary casing should always be included in the items covering for drilling in a contract document. If the surface deposits are substantial, i.e. greater than 2 m in thickness, then rotary open hole drilling, although slower than percussive drilling, may prove to be more effective. In addition to the slower rate of drilling, rotary rigs are, with a few exceptions, not track mounted, therefore they are less manœuvrable than rotary percussive rigs. The typical cost per metre for drilling open hole boreholes with rotary percussive rigs is considerably less than with non-percussive rigs.

Injection plant. The type of grout injection plant required for a particular site will depend on the extent of the area to be grouted, the estimated quantity of grout to be injected and the time available to complete the work. In general the plant required comprises a mixer, measuring hopper, or weigh batcher, water tank, agitator, grout tubes and pipes, a pump capable of injecting grout at a specified pressure and a borehole sealing device equipped with a pressure gauge and relief valve. The plant should be assembled so as to obviate spillage into existing drains or on to roadways. If the estimate of grout required is between say 100 and 500 t a single drum or double drum mixer is adequate. The output from this type of mixer generally averages, according to site conditions, from 20 to 50 t/day. Where the estimate is between 500 and 1000 t the minimum requirements for mixing plant would be a double drum machine capable of averaging 50 t/day. For grout quantities in excess of 1000 t mixing and batching plant capable of a daily output of 100 t or more should be used. Combined mixing and batching plant such as the large Cumflow machines, incorporating powered buckets and shovels, can deal comfortably with large grout outputs. Portable cement silos of approximately 30 t capacity are normally used in conjunction with this type of mixing plant.

The decision as to the size of plant required for a site can also be influenced by the number of boreholes required. It should be borne in mind that the daily grout output should not be such as to fill more boreholes than can be drilled in the same period, otherwise plant standing time will occur. Plant layout should be carefully planned to enable the normal sequence of operation to proceed without interruption. Materials are placed in the batching bucket, weighed and put into the mixer and water added. The mixed constituents are then transferred by chute into an agitator and then pumped to the boreholes through pipes and grout tubes. The accuracy of the bucket and cement silo weights should be checked daily. With regard to the smaller single and double drum mixers a measuring box is generally used to ensure reasonable accuracy and consistency of the weight of the grout materials fed into the mixer. Where small quantities of grout, e.g. less than 100 t are required for the filling of, say, localized natural cavities in limestone or large rock fissures in foundation areas, the plant may comprise a simple mixer, agitator, delivery pump and grout pipes.

Ram and centrifugal pumps are commonly used for injecting grout. The type of pump selected for a particular site will depend on the depth of the boreholes, the distance from the mixing and batching plant to the grouting location, the volume of grout being delivered and the pressure required.

Materials and workmanship

Grout mixes. Grout mixes required to fill disused underground mineral workings may comprise a cement/p.f.a. mix, a sand/cement/p.f.a. mix or a sand/cement mix. As a grout constituent, p.f.a. is ideal; it is a pozzolan and therefore produces a stable durable grout when combined with cement. The particle shape of p.f.a. is spherical, enabling grout to be pumped more easily than a grout composed of sand and cement. The p.f.a., which should be conditioned hopper ash and not lagoon ash p.f.a., is preferable to sand in grout mixes on both economical and ecological grounds. Montmorillonite clays such as bentonite may be used as an additive to grout mixes to increase the viscosity of the grout, particularly where the mine workings are inclined.

Grout strength and mix design. The strength of the grout mix will vary according to the requirements of each individual site. The strength will depend on factors such as the nature of the superincumbent strata and the size of the structure to be supported at the surface. Generally the grout mix as a ratio of p.f.a. to cement may vary from 6:1 to 10:1. The compressive strength of the grout at these ratios will vary from 1 to 4 MN/m^2. The water content of the grout mix is a critical factor in the determination of grout strength. For cavity grouting with a cement/p.f.a. mix the water content should not exceed 0·4 times the total weight of the solid constituents.

When p.f.a. is used as a grout constituent it should be covered during periods of wet weather. Regular checks should be made to ascertain the water content of the p.f.a. and, if necessary, the water ratio should be adjusted accordingly. For the majority of cases ordinary Portland cement is adequate for grout mixes but for flooded workings in sulphate bearing strata, it may be necessary to use sulphate-resisting cement.

Limiting pressures. For the grouting of the old shallow mine workings low pressure infill grouting only should be undertaken. At depths of 10 m or less pressure should not exceed 100 kN/m^2. At depths of over 10 m, pressure should not normally exceed 200 kN/m^2. High pressure infill grouting is not recommended at shallow depths as uplift at ground surface will occur. For perimeter grouting, pressurization is not required as the purpose of the grout injected is to form an artificial wall or barrier to contain the infill grout. Where existing structures are in close proximity to proposed grouting areas then it is essential that adequate supervision by an experienced engineer is provided during the pressurization stage of the grouting operation. If necessary, observations by precise level should be taken at selected points prior to and during the grout pressurization period to ensure that uplift of the structures does not occur.

Records and testing. Accurate records should be kept showing the location, depth and injection sequence of all grout boreholes, the quantity of grout taken by each and the maximum pressure of injection. Records of materials delivered to the site should be kept and checked regularly against quantities used. Daily summaries of the amounts of grout injected, materials used and borehole depths should be maintained by the contractor and site supervisory staff. It is advantageous to the contractor and employer to maintain progress graphs showing the actual daily tonnage of grout injected and borehole metrage achieved against the programmed daily output to ensure that progress is satisfactory. The adequacy of the grouting works should be tested either by water tests or the taking of core samples. Where water tests are undertaken boreholes not exceeding 70 mm in diameter may be drilled at selected locations to intercept the grouted mine workings. Water is injected into the boreholes and a pressure of between 100 and 200 kN/m^2 will require to be held, depending on the depth of the borehole, for approximately three minutes. If there is no appreciable fall, i.e. 10% in pressure, during this period, the test can be deemed satisfactory. The most effective way of testing grouted areas is by taking borehole core samples. This is done by rotary drilling and producing cores from the grouted materials. The major disadvantage of this method is the high cost of obtaining cores.

Application to old workings
Groundwater. Generally groundwater is not used as a major source of domestic water supply in mining areas. The Bunter and Keuper Sandstones of the Trias are important aquifers. The Permian Limestone yields substantial quantities of water and the sandstone of the Carboniferous Coal Measures has been drilled for water. Yields in the Coal Measures have been much less than abstracted through mine drainage. Groundwater may or may not be present in old mine workings where grouting is required. This will depend on the degree of permeability of the overlying strata and whether the workings are steeply inclined or are situated near the centres of

anticlines or synclines. Groundwater does not normally cause major difficulties to either grout drilling or the grout injection operation. The setting properties of the grout will not be affected by groundwater providing the grout is injected through tubes to the base of the workings and thereafter by progressive withdrawal of the tubes as the cavities and boreholes are filled. Rapid setting and hardening cements are not recommended for grouting. Groundwater in mine workings may contain sulphates in solutions which attack the solidified grout. In such cases a sulphate-resisting cement should be used in place of ordinary Portland cement.

Limitation of depths. The limit of depth at which grouting is undertaken must always be assessed in relation to the level of the base of the foundations of the structure to be supported including the base of driven or bored piles. As a general rule grouting of old mine workings is undertaken beneath the base of the foundations of a structure at depths of 15 m or less or when the rock cover is less than 10 times the thickness of the seam extracted, whichever is the greater. In the case of motorway construction or major road improvement schemes grouting would not normally be undertaken beneath the carriageways at depths in excess of 10 m below formation level except in the case of associated structures.

Drilling patterns. The drilling pattern for the grouting of old mine workings should always comprise the minimum number of boreholes required to stabilize and support the strata between the roof of the old workings and the structure or highway to be supported. Obviously a 100% fill of voids would be achieved if the boreholes were spaced at very close intervals but in practice the spacing must be determined to ensure that the structure or surface will be adequately supported at the minimum cost. With regard to the extent of the area to be grouted, the periphery should extend beyond the outline of the structure by a distance of not less than 0·75 times the depth of the seam from the base of the foundations to be supported. The pattern of boreholes required for a particular site will not vary appreciably and will normally comprise perimeter and infill boreholes. Perimeter boreholes will be drilled at centres which may vary from 1·5 m to 2·25 m and infill boreholes at 3 m centres. It is preferable to stagger alternate rows of the infill boreholes to form a 'diamond' rather than a square pattern.

Grout sequences. When the drilling pattern has been selected the position of the grout holes can be set out to cover the area to be stabilized. Perimeter borehole drilling commences and the injection of grout can proceed when sufficient boreholes have been drilled in advance. Depending on the depth of the boreholes, grouting should not be undertaken within 10 m of the borehole or boreholes being drilled. The grouting of the perimeter boreholes comprises the first stage of the grouting operation. The perimeter boreholes are grouted initially to provide a barrier to contain the infill grout. This operation is particularly important when the old mine workings are inclined. Where the workings are inclined or the cavities to be grouted are large, pea gravel may be introduced into the perimeter boreholes. In certain cases bentonite may be used as an additive to decrease the flow rate of the grout. Normally the proportion of bentonite required for perimeter grouting is 10% of the cement constituent. Perimeter boreholes do not normally require pressurizing after filling. Drilling of the infill boreholes may commence prior to completion of grouting of the perimeter boreholes but as stated previously, drilling should not be undertaken within 10 m of the holes being grouted. The second stage of the grouting operation is to fill the infill boreholes. Grouting should commence from the dip side of the seam and normally not more than two boreholes are injected at the same time. The grout should be introduced, preferably at the base of the borehole, by means of grout pipes or tubes. Grout tubes should be hollow metal screw-jointed rods or, for ease of use, Alkathene tubing may be selected. Grout pipes or tubes are inserted into the borehole to ensure that there are no blockages in the hole and that the grout can reach its true destination which is the cavity or broken ground. When the grout emerges at ground surface from the borehole, the pipe or tube is gradually withdrawn. After each infill borehole fills with grout under delivery pressure, the hole is sealed and then pressurized.

Design considerations (pressures, cavitation of overlying strata etc.). When considering the technique required to grout a particular site, due consideration must be given to the nature of the strata overlying the old mine workings and the type of construction of the proposed buildings. If a thick bed of massive sandstone is the predominant stratum between the old workings and the surface then the upward migration of cavities will be severely restricted. However, it is imperative for the sandstone to be adequately supported by a grout of a strength capable of taking the load of the superincumbent rock and the structure. If the predominant material is shale then cavities will migrate towards the surface and voids may be present well above the level of the old workings. Beneath the voids loose shale or mudstone will be present and the grout mix for this type of situation should be strong enough to support the overlying strata and structure, as stated above, and also be capable of penetrating and stabilizing the loose shale. If the procedures adopted in the section on grout sequences are carried out then any cavities or broken ground overlying old mine workings will be fully grouted and will not be a hazard to the surface development. Depending on the type of structure to be supported and the method adopted for extraction of the coal in the past, it may be possible to utilize a partial system of grouting. This method does not ensure that the mine workings are fully grouted but provides columns or cones of grout to support the overlying strata at fairly regular intervals. A detailed site investigation is essential prior to carrying out this type of operation. The disadvantage of this method of grouting is that the spacing of the columns of grout cannot be guaranteed unless the boreholes are relatively closely spaced. This would of course nullify to a large extent any saving in cost on full grouting methods.

Application to fissures

Grouting of minor fissures, which vary in width up to 5 mm, is undertaken only by chemical grouting and not by grout constituents employed for the filling of the large cavities resulting from the extraction of coal or other minerals. Large natural fissures may be present in sandstones and limestones beneath the foundations of structures. Stabilization of the rock, particularly when the strata are inclined, can be achieved by the use of p.f.a./cement grout. In active mining areas, natural fissures may be widened by tensile strains induced by the extraction of the minerals. The layout of the borehole pattern will be similar to that utilized for the grouting of mine workings but the boreholes should be inclined at approximately 2:1 from the vertical. Therefore with surface spacing of the boreholes at 3 m and borehole depths of 6 m, all major fissures within the affected area should be intersected with a staggered pattern of boreholes. Grout can be injected directly into the open fissures where practicable. For fissure grouting, final pressures should not exceed 100 kN/m^2. Testing of the grouting may be undertaken as described earlier. Where large fissures occur, but do not affect structures, e.g. across school playing-fields, grouting is not normally undertaken. The fissures, depending on their width and depth, may be benched out and then filled to benching level with shale or lean concrete and then capped with a lightly reinforced concrete raft to provide a permanent remedy.

EXCAVATION AND FILLING
Shallow cavities

Ground stabilization in areas where old mining operations have taken place, or where natural cavities are present, can be undertaken by full excavation of the overlying strata to the floor of the old workings, or by partial excavation and compaction. Excavation and filling are employed where grouting would be uneconomical in both expense and time. Shallow cavities can be defined as those cavities which occur at depths of between 1 and 10 m from the surface.

Man-made cavities. Coal and other minerals have been exploited by man over several centuries from the outcrop of the minerals to depths which have increased considerably with the advancement of scientific and technological knowledge. As far as coal is concerned, legislation imposes on the National Coal Board a statutory duty to keep accurate records and plans of their mine workings. This was not the situation when many mines were worked at very shallow depths, particularly during the 18th and 19th centuries. Many unrecorded workings are believed to exist, and if surface development is contemplated, then it is imperative to carry out a detailed site investigation if the geological information indicates that minerals exist, or have existed, in close proximity to the surface. During construction of the M1, M62 and M621 motorways in Yorkshire, exploratory drilling revealed the presence of uncharted mine workings in close proximity to the surface. The coal seams were worked by the pillar and stall method, i.e. tunnels driven in the seam from the outcrop with a pillar of coal left to support the tunnel. Shallow mine shafts were also sunk for ventilation purposes. Due to the shallow depths of the workings, the roof of the tunnels deteriorate and collapse, or the pillars may crush with the passage of time and cavities may eventually migrate to the surface. Many of the tunnels remain intact, however, and present a hazard to surface development due to their shallow depth. This type of mining has been described in detail in Chapter 1. Where cavities due to man-made workings are located, then stabilization by excavation and backfilling should be carried out if the economies of the operation justify the method in preference to grouting. For structures, excavation is not generally employed where old mine workings are present at depths in excess of 5 m. In the case of motorways or highways under construction, excavation to remove old workings may be undertaken to a maximum depth of 10 m. Where excavation is undertaken, the area must be carefully backfilled and compacted. Backfilling and compaction should be in accordance with the relevant Class defined in the DoE Specification of Roads and Bridge Works. Where the cover between old mine workings and the top of the embankment is more than 10 m, full excavation or grouting should not be undertaken. At the toes of the embankment, however, the cover may be considerably less than 10 m and remedial works may be required. Stabilization in this case can be achieved by excavating a trench 5 m to 8 m in width along the line of the toe of the embankment. The width of the trench will depend on the depth of the mine workings. Excavation will be continued until the depth of cover, which will depend on the nature of the strata, is a minimum of 5 m. The excavated trench will be backfilled and compacted as described above. Geological faults may be present along the route of the motorway or highway and where the surface position of a major fault, i.e. with a displacement of more than 3 m, is located, partial excavation in the form of back-benching will be required. Back-benching will be undertaken by excavating the undisturbed ground to a depth of 2–3 m at the surface position of the fault, and benching back to ground or formation level over a distance of not more than 15 m on both sides of the fault. Selected backfilling should then be provided in 150 mm layers to a 5% air voids specification.

Bell pits. During the 18th and early 19th centuries, coal, and in particular ironstone, was mined at a shallow depth from bell pits. The name is derived from the shape of the extracted area. A shaft was sunk to the mineral to be exploited, and then worked as far as possible around the base of the shaft, roughly in a circle, until there was a danger of collapse, as no roof supports were used. A second shaft was sunk as close to the previous workings as possible and the material excavated was used to fill the prior excavation. This process was continued until the depth of the material increased to such an extent that it was no longer economic to work by this method. The bell-shaped areas of excavation, as located by site investigation and surface development, are generally filled with clay or shale, or a mixture of both materials. Many old coal and ironstone bell pits were located during the construction of the M1 motorway in Yorkshire. The frequency of these shafts is particularly concentrated in the Tankersley and Thorpe Hesley districts of South Yorkshire. Cavities are not normally present in bell pits; the fill material is, however, usually loose and uncompacted. Where bell pits are present at depths of 10 m or less, stability may be achieved by subjecting the area to heavy vibratory compaction; the weight of the compacting roller depends on the strength of the substrata. Site investigation of the area is essential, however, before determining the method of stabilization to be adopted. Generally, compaction alone can be utilized only in a small number of cases, but a combination of partial extraction and compaction can produce the degree of stability required.

Natural cavities. Natural cavities may be present in limestone and sandstone at very shallow depths from the surface. The formation of this type of cavity has been described in detail in Chapter 1. The presence of such cavities can be determined only by inspecting geological maps and records of the area under review, assessing local knowledge of the district and carrying out a detailed site investigation. Before excavation commences, the extent and shape of the cavity should be ascertained as accurately as possible by drilling, which should be more economic than the use of geophysical methods, to enable the contractor to remove the overlying strata to an approximate depth of 3 m without danger to his plant operators and damage to his machines. Depending on the nature of the overburden, it should be removed by either ripping or blasting. Backfilling and compaction of the former cavitated area should be undertaken as described in the section on man-made cavities.

Fissures. Generally, major fissures in limestone and sandstone located during excavation for structures and highways, are dealt with by grouting, but where they occur across, say, school playing fields, car parks, or recreational areas, remedial works may be undertaken initially by filling the fissure with a granular type material. Benching along the line of the fissure should then be undertaken at a width not exceeding 1 m on both sides of the fissure. The fissure should then be capped with a lightly reinforced concrete slab not less than 100 mm thick, the capping extending to the full width of the benching.

Adits and tunnels

The term adit refers to horizontal or inclined shafts driven from the surface either in the coal or mineral seam, or through other strata to intercept the seam. Where development is contemplated, in areas where outcrops of coal or other mineral seams occur, detailed investigation of National Coal Board and other mineral working plans should be carried out. Geological plans of the area should be consulted, and if mine adits are believed to be present within the development area, site investigation should be undertaken. If there are no visible signs of adits on the surface, then excavation by means of trial trenches should be carried out at their approximate locations. The exploratory area should extend for a minimum distance of 30 m in all directions from the assumed surface position of the adit. When located, partial excavation should be undertaken to ascertain whether the adit has been backfilled, or merely sealed off at the entrance. If the adit has been walled off, then it should be opened out from the surface by excavation to a minimum depth of 5 m. Additional excavation or grouting beyond this depth will be dependent on the nature of the overlying strata and the size of the adit. The excavation should be backfilled and compacted as described in the section on man-made cavities, but if the adit functions as a drainage channel for the mine, adequate drainage facilities will have to be provided prior to backfilling. If development is contemplated above the adit location, then a reinforced concrete slab should be provided. In the case of motorways, or major improvement schemes, horizontal or inclined adits located in cuttings should be drained as necessary and walled up at the cutting face with lean or porous concrete, the exposed face of the concrete being set back and left rough to receive 200 mm to 300 mm of topsoil. In rock cuttings face walling of masonry using rock similar to that of the adjoining rock may be used in lieu of concrete. Where adits or tunnels encountered in the sides of cuttings are of such inclination that they extend below formation level and water would drain into them, a 3 m thick clay plug should be provided behind the face walling. The plug should be supported with stone filling in the adit or tunnel and then sealed with a mortared face wall 300 mm thick.

Shafts

The procedure and investigation for locating abandoned vertical mine shafts will normally be undertaken as described in the previous section for adits. If there are no visible signs of the shaft on the surface, excavation by trenching is undertaken. Shafts may appear to be filled but verification should be obtained by drilling boreholes both inside and outside the shaft to determine the extent of the filling (staging may be present at a shallow depth from the surface) and the nature of the ground immediately surrounding the shaft. Suitable safety precautions should be taken prior to commencing drilling. e.g. the drilling rig should stand on beams or baulks of adequate size to span the shaft in the event of collapse of the shaft fill. Normally one hole is drilled down the centre of the shaft to a depth not exceeding 30 m with three holes around the perimeter of the shaft drilled to rockhead. In the case of abandoned coal mine shafts, approval of the treatment of the shafts must be obtained from the National Coal Board. Where shafts are found to be open, the bottom 15 m of the shaft should be filled with boulders approximately 600 mm in size, to allow any mine water to percolate without interruption, and to prevent loss of fill into the mine workings. The remainder of the shaft should be filled with clean rock fill or other type of material free from fines and organic material. All shafts shall normally be covered with a square doubly reinforced concrete slab, where possible, constructed at rockhead. The size of the slab shall be approximately twice the diameter of the shaft. In exceptional cases, e.g. where founding level cannot be undertaken at rockhead, the size of the slab may be increased to $2\frac{1}{2}$ times the diameter of the shaft. In such cases, grouting of the shaft fill may be required, but this will be dependent on the information provided by the shaft drilling. The responsibility for treating abandoned mine shafts which exist within housing or industrial estates lies with the surface developer. The treatment proposed, i.e. filling material, type of slab and reinforcement, will require the approval of the National Coal Board. The National Coal Board's liability in respect of abandoned coal mine shafts is contained in Sub-Section 1 of Section 151 of the Mines and Quarries Act, 1954, which limits its liability to the provision of an effective enclosure, barrier, plug or other device to prevent any person accidentally falling down the shafts.

REFERENCES

Bell, F. G. (Ed.) (1975). *Methods of treatment of unstable ground.* London: Newnes-Butterworth.

Cambefort, H. (1977). The principles and applications of grouting. *Q. Jnl Engng Geol.* **10**, 57–95.

Institution of Civil Engineers (1963). *Symposium on grouts and drilling muds in engineering practice.* London: Butterworth.

Chapter 4
Structures

SUSCEPTIBILITY OF STRUCTURES TO PREDICTED GROUND DISPLACEMENTS

Relationship between ground displacements and deformations of a structure

Ground subsidence is a phenomenon which may be beyond the control of the engineer responsible for the design of a structure. Displacements of the ground surface directly attributable to active mining may be predicted with reasonable accuracy but many forms of ground subsidence cause severe unpredictable local ground displacements. A structure will attempt to resist the loss of local support. If it possesses sufficient strength and stiffness, its deformations will not be directly related to the magnitude and pattern of large local ground displacements. However, the effects of such ground displacements can be disastrous in the case of a structure which possesses insufficient strength and stiffness and is largely dependent on the support provided by the ground.

Traditional low-rise load-bearing brickwork buildings founded on unreinforced concrete continuous footings are readily damaged by local ground subsidence and the deformations of such buildings conform to the ground displacement profile.

High-rise reinforced concrete shearwall buildings generally possess sufficient strength and stiffness to ensure that the structure 'bridges' the local removal of ground support until it can be restored by remedial works, or by the passage of the subsidence wave. Thus, either the structure or its ground support may be the dominant element or they may be balanced.

Considerations of structure–soil interaction

Structure–soil interaction can be directly caused by the imposition of a building on a yielding foundation, or indirectly caused by ground displacements unrelated to the construction of the building. The interaction resulting from mining subsidence can be of greater severity than that which would be caused by the weight of the structure being placed on a yielding foundation.

Structure–soil interaction is an inevitable mechanism which occurs with statically indeterminate structures and yielding foundation systems, the effects of which can be accommodated or minimized by deliberate design and construction techniques provided that ground displacements can be foreseen and their type and magnitude predicted. If, however, the ground subsidence occurs as an unexpected event during the life of a structure, then the initial load distribution for which the structure was designed may be dramatically changed and secondary stresses may be induced which can adversely affect the performance of such a structure. Secondary stresses induced in the structural elements are not the sole consideration, since structural deformations may provide a more serious problem in cases where the induced stresses are tolerable. It is important to appreciate the interdependence of the foundation system and the superstructure, which should be designed as balanced and integrated systems. It is equally important that the design engineer should be involved in the design process at as early a stage as possible since adverse effects of structure–soil interaction can be minimized by appropriate design and construction techniques.

Reference may be made to a report on structure–soil interaction prepared by the Institution of Structural Engineers (1977) for detailed information on this complex subject.

Table 1

Class of structure	Type of structure	Limiting angular distortion
1	Rigid	Not applicable: tilt is criterion
2	Statically determinate steel and timber structures	1/100 to 1/200
3	Statically indeterminate steel and reinforced concrete framed structures, load-bearing reinforced brickwork buildings, all founded on reinforced concrete continuous and slab foundations	1/200 to 1/300
4	As class 3, but not satisfying one of the stated conditions	1/300 to 1/500
5	Precast concrete large panel structures	1/500 to 1/700

Allowable differential settlements

Allowable differential settlements have been defined by various investigators on the basis of their studies of the behaviour of different types of structure. Skempton and Macdonald (1956) investigated the behaviour of about one hundred buildings of reinforced concrete and steel frame construction and also of traditional load-bearing wall construction. The criterion of damage used by Skempton and Macdonald was the angular distortion which is defined as the ratio of the differential settlement and the distance between any two points on a structure with the component due solely to tilting eliminated. Skempton and Macdonald concluded that damage to the elements of structural frames is likely to occur if the angular distortion exceeds about 1/150. They also concluded that structural cracks will develop in the load-bearing walls of traditional buildings and in the panel walls of frame buildings if the angular distortion exceeds 1/300.

Bjerrum (1963) recommended that the safe limiting value for angular distortion where structural cracks are not permissible is 1/500 but agreed with Skempton and Macdonald that the limiting value for angular distortion where structural damage to buildings may be expected is 1/150. Starzewski (1974) drew attention to a comprehensive study of the performances of seventy structures. This study was published by the Institute of Building Technology, Warsaw, and draws the conclusions concerning building performances shown in Table 1.

Since ground displacements caused by local ground subsidence can cause angular distortions of as much as 1/50 in traditional low-rise brickwork structures then it is apparent that special consideration must be given to structural deformations in districts where ground subsidence may be expected and appropriate measures taken in the design and construction of such buildings. A better definition of settlement limits is given by relative deflexion which is the ratio of the maximum displacement relative to the straight line between any two points on a structure to the distance between the two points. Polshin and Tokar 1957 used this criterion and also introduced two further criteria which are of fundamental importance, namely the length to height ratio of a wall and the concept of limiting tensile strain before the development of cracks. These criteria have been discussed in depth by Burland and Wroth (1975).

Allowable tilt

The definition of tilt is the rigid body rotation of a complete structure or a well-defined portion of a structure. Chimneys, towers, and masts may be placed in this category but flexural deformations of tall slender structure render difficult the measurements of tilt since the behaviour is no longer that of a rigid body. There is little evidence available to permit limits of tilt to be stated with confidence. 1/250 may be considered a limiting value for tilt of tall slender structures provided stress analyses establish the acceptability of the induced stresses due to lack of verticality.

Effects of tensile and compressive horizontal strains

Ground subsidence is generally accompanied by horizontal ground strains (see Chapter 2). In cases other than active mining, the horizontal strains which develop, particularly those where local subsidence is experienced, can be estimated only from experience and the advice of a specialist should be sought.

The multiplicity of structures within conurbations affected by ground subsidence necessitates that they be considered individually and general comments only will be made in this section.

The horizontal ground strains may be compressive or tensile and the latter can cause significant damage to traditional brickwork buildings supported by unreinforced concrete footings. Equally, structures which have a low resistance to compressive forces can be damaged by relatively large horizontal compressive strains. The structural strains are induced by the drag which is imposed on the external surfaces of the bases of foundations by frictional forces. The horizontal displacements of the ground relative to the foundation—ground interface cause a lengthening or shortening of the foundation depending on the type and magnitude of ground strain and on the resistance of the foundation to horizontal displacements. The mechanics of relative displacement and development of frictional forces may be demonstrated by consideration of the design of a shallow reinforced concrete slab foundation supporting a traditional low-rise brickwork building.

The relationship of the basic components of the problem is shown in Fig. 8 and it will be assumed that horizontal tensile strains are being induced by ground subsidence. This idealized case of balanced ground displacements about the centre-line of the slab foundation results in a triangular distribution of relative displacements. The design of the base slab should therefore take into consideration normal structural requirements combined with the effect of the tensile ground strains on the overall stresses in the reinforcement. In the idealized case it is assumed that the drag forces are sufficient to ensure that the horizontal displacements of the slab and ground are equal. Relative movements or slip between slab and ground will, however, occur in practice. The drag forces are proportional to the weight of the structure and the frictional restraint ratio at the foundation–ground interface. It is desirable, therefore, to construct a horizontal plane slab foundation and form a separation layer. The frictional restraint ratio may then be expected to vary from 0·50 to 0·70 and a value of 0·67 is commonly adopted. For the idealized case under consideration the predicted maximum horizontal force on the slab foundation then becomes one third of the total weight of the structure.

It is assumed for the purposes of distribution of reinforcement that the drag forces have a distribution similar to that shown in Fig. 8. In order to resist the effects of horizontal tensile ground strains the area of steel reinforcement should be determined on the basis of the predicted maximum horizontal force in the slab foundation and the permissible stress in the steel reinforcement. A similar but reversed stress situation occurs with horizontal compressive ground strains and the design engineer must ensure that the compressive resistance of the structure is not exceeded. Although the example of a continuous slab foundation was selected for ease of understanding the same basic concepts apply to structures founded on separate foundation elements. In view of the serious adverse effects of compressive and tensile ground strains on structures it is essential that the ground drag be minimized and piled foundations should not be used in conditions where the dowelling action of piles in moving ground may create the worst possible condition of restraint.

Avoidance of statically indeterminate structures

Statically indeterminate structures should be avoided in cases where the predicted ground movements are significant since the initial load system is completely changed by the interactive effects and also accurate structural analyses in such cases are difficult, if not impossible.

Avoidance of basement accommodation

The construction of basements in ground which will probably be subjected to horizontal and vertical displacements due to subsidence is not recommended. The retaining walls of basements will be subjected to significant additional lateral ground pressures in cases where horizontal compressive strains are experienced. The drag forces on the vertical surfaces of basement walls augment the drag forces on the base slab in cases where horizontal tensile strains are experienced. As a general rule it is preferable to superimpose structures on the surface of ground which will be subjected to subsidence since this approach results in a simplification of design and construction techniques. Shallow foundations also facilitate the execution of

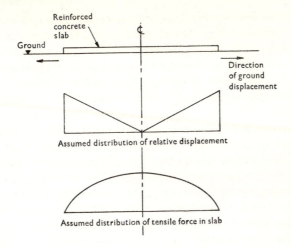

Fig. 8

any necessary remedial works. Reference should be made to the report on design and construction of deep basements published by the Institution of Structural Engineers (1975).

DESIGN OF FOUNDATIONS TO ELIMINATE UNACCEPTABLE EFFECTS OF SURFACE MOVEMENT
Transmitting loads to below level of natural cavities or old workings
The loads from structures may be taken below natural cavities or old workings by the following methods:

(a) spread foundations located below the cavities;
(b) piles terminating below the cavities.

Spread foundations may take the form of piers, bases or rafts constructed either in individual excavations or where columns or walls are closely spaced in a large open excavation.

Piles should not be used unless subsidence is virtually complete, as large movements of the ground will impose high horizontal loads on piles which they may be unable to resist in shear. However, when subsidence is virtually complete but infilled voids remain, piles provide a useful method to transfer the load to a lower stratum below the zone of subsidence. The possibility of further small movements of the ground usually remains and the piles should be designed to withstand additional loading resulting from downdrag of the surrounding ground. Large-diameter bored piles with a minimum shaft size of approximately 1 m are suitable; they may be heavily reinforced to increase their resistance to horizontal forces and are capable of supporting large vertical loads so reducing the total number of piles required. This keeps to a minimum the costly operation of drilling through collapsed soil or rock in the zone of subsidence. The piles will usually be founded in sockets cut into the bedrock below the subsidence zone. With a 1 m minimum diameter shaft, a man may enter to inspect the work or use a jackhammer to loosen rock to facilitate mechanical boring. The shaft of a pile should be sleeved through cavities, a permanent light gauge steel lining tube inside the temporary casing being commonly used. Even when a cavity is not encountered in a pile bore, it is advisable to use full length permanent lining, as the lateral pressure of wet concrete may be sufficient to break through the side of the bore into a cavity. This may result in damage to the shaft of the pile as the concrete above the level of the cavity may arch across the pile and not fall down to replace concrete lost to the cavity. Such a defect is unlikely to be detected and could lead to structural failure.

Lateral and downdrag forces on a pile may be reduced by surrounding the pile shaft with bentonite, or by coating the permanent lining with a thick layer of soft bitumen. When piles have to be constructed through water the concrete should be placed using a tremie pipe and hopper to avoid loss of cement and fines. When the head of water is not great, a plug of concrete of sufficient depth to resist the hydrostatic pressure may be formed under water in the permanent lining tube. When the concrete has hardened the lining tube can then be pumped dry and the shaft of the pile constructed in the normal way.

Grouting of old workings

The plant and the materials required for the infilling of cavities are described in Chapter 3 together with the application of the technique to old workings. When the grouting is complete conventional foundations are constructed to suit the nature of the overburden. It is good practice to keep the foundations as high as possible above the consolidated cavities to take advantage of the dispersion of load, so avoiding concentration of stress on the grouted rock. It is bad practice to use piles over grouted cavities unless an adequate thickness of sound rock exists between the toes of the piles and the infilled cavities.

A natural arch of competent rock is particularly desirable over old mine workings as loose stowage and soft seat earth may not have been strengthened by the grouting work. Where there is doubt as to whether the consolidation work will be fully effective in preventing further collapse of the ground above infilled workings, a structure may be further protected by the provision of jacking facilities to correct the level of the building if this proves to be necessary.

Three point support systems

Subsidence may be accepted by structures without distress if they are sufficiently rigid to be supported on only three points. Subsidence of any one point will cause the structure to tilt about an axis drawn through the other two points. The principles of such a design are as follows:

(a) foundations are made as small and stiff as possible to reduce and resist the bending moments, and extension or compression of the ground during subsidence;

(b) ground pressures are kept as high as the subsoil conditions permit in order that pressure redistribution may take place under the foundations;

(c) the structure is located on a sliding layer to reduce friction;

(d) a strong rigid superstructure is required;

(e) jacking facilities are provided at each of the three foundations for relevelling after subsidence.

DESIGN OF FOUNDATIONS TO MINIMIZE THE EFFECTS OF SURFACE MOVEMENT ON THE SUPERSTRUCTURE

Raft foundations

The main principles to be followed in the design of raft foundations are first to make them as shallow as possible, so that compressive strains in the ground can take place beneath them instead of transmitting direct compressive forces to their edges, and, secondly, to construct the rafts on a membrane so that they slide as ground movements occur beneath them. It is then necessary only to provide enough reinforcement in the rafts to resist tensile and compressive stresses induced by friction in the membrane in addition to the normal requirements of structural design.

In the case of heavy structures it is desirable to adopt the highest permissible bearing pressures so that the plan dimensions of the raft are as small as practicable. By this means the total horizontal tensile and compressive forces acting on the underside of the raft are kept to a minimum, and the length of raft acting as a cantilever at the 'hogging' stage, or as a beam at the 'sagging' stage of the subsidence are also at a minimum. When the design bearing pressure is increased for the above reasons so that the factor of safety relative to ultimate bearing capacity is significantly reduced, the consolidation settlement may be severe if the ground is compressible, but the magnitude of the consolidation settlements is likely to be small in relation to the movements from mining subsidence. Such considerations need to be evaluated in determining the extent to which the factor of safety can be reduced.

Raft foundations may take the form of simple slabs, or slabs stiffened locally by upstand beams for light structures. Rafts for heavy structures may be thick slabs, fully stiffened by a rectangular grid of upstand beams acting compositely with a ground floor slab. In the case of light structures such as dwelling houses, it is impracticable to vary the size of raft from the plan area of the building and, on large estates, the additional cost of such precautions may not be viable in relation to the prospect of damage being caused to a limited number of dwellings. In such cases ordinary strip footings may be used but similar principles should be followed to prevent ground strains being transmitted to the foundations. Consequently trench fill methods of construction should not be adopted. In such cases it is desirable that the ground floor slab be designed as a suspended slab with sufficient reinforcement to resist horizontal strains which could cause differential lateral displacement of individual footings with consequent damage to the superstructure.

Provision for strains in soil below the foundation
Irrespective of the type of raft foundation adopted it is important to ensure that the horizontal forces transmitted to the raft from the movements of the ground are minimal. With this objective, the raft should be constructed on a membrane laid over a granular base and the underside of the raft should be flat. The following construction principles are recommended.

(a) Light structures—A layer of compacted sand or pulverized fuel ash or other suitable granular material 150 mm thick laid on a prepared formation and covered with polyethylene sheeting on which the foundation is constructed.

(b) Heavy structures—A layer of blinding concrete approximately 50 mm thick is laid on a prepared formation and covered with polyethylene sheeting and a layer of compacted sand or pulverized fuel ash or other suitable granular material 150 mm thick. Polyethylene sheeting is laid on the sand and the construction is given protection by an unreinforced concrete slab 100 mm thick on which the foundation is constructed.

Provision for compressive forces in soil around substructure
Trenches excavated around both new and existing buildings may be used to

reduce the compressive forces transmitted to the foundations. Trenches placed as near to the structure as possible and to a depth just below the underside of the foundations, absorb strain by allowing the ground to compress unconsolidated backfill in the trench. For practical reasons the trench has to be filled with a material strong enough to support the sides but more compressible than the surrounding soil so that crushing of the filling may take place. Coke has been used for this purpose but consideration may be given to the use of expanded polystyrene in slab form. The trenches are covered by paving slabs or by other suitable means. Foundations constructed as a number of small units can be given more protection by these methods than foundations covering a large area of the ground. In the case of multi-storey buildings, special design consideration has to be given to the means of catering for wind shear. There is a conflict of design requirements between the wind resistance and mining subsidence resistance of the structure and it may be necessary to accept a reduced factor of safety in respect of resistance to wind shear.

Implications of tilting on structures

The passage of a wave of subsidence will cause a structure to tilt before it subsides to its final position. The siting of a structure in relation to the wave front is important; there will be a greater amount of ground support where the long axis of the structure is parallel to the wave front than in the case of a building with its length at right angles to the wave front. Structures should not be sited within 15 m of known geological faults, as subsidence and tilting is likely to be severe near fault planes. If the structure is isolated, tilting may be inconvenient causing malfunction of lifts, drainage and other services but it is unlikely to cause structural damage. Where several structures are built in close proximity or with only a nominal gap between (as at an expansion joint), significant structural damage can be caused as the angle of tilt is unlikely to be uniform and large compressive forces will develop between the buildings either at the top or at the bottom, depending on the angles of tilt.

The National Coal Board (1975) give an example of a gradient of 1 in 250 occurring as a result of extracting a seam 1 m thick at a depth of 600 m. This tilt is unlikely to cause any significant reaction in a structure due to redistribution of weight. A chimney whose height is 40 m would be 160 mm out of vertical, which is tolerable as a temporary state under conditions which will lead to its return to the near vertical when subsidence is complete. With the same tilt an eleven storey block of flats having a height of about 27·5 m would be 110 mm out of vertical. This is unlikely to affect the stability of the building but could affect the operation of the lifts. This particular problem can be allowed for by constructing enlarged lift shafts with provision for adjustment of the guides to true verticality. A tilt of this order is undesirable as a permanent feature, although tower blocks are not built to an accuracy of less than 25 mm.

At extra cost, structures may be provided with facilities for jacking them level, during or after the passage of the subsidence wave. This can enable sites to be developed which would otherwise be considered unsuitable for development. Equally it can permit buildings with a lower tolerance to disturbance, such as brick or masonry structures, to be considered for sites which may be disturbed by subsidence. Such provisions are dealt with in a later section.

When a gap is left between buildings or sections of a building and for convenience of construction the gap is filled with a compressible material, care should be taken to ensure that the material selected is sufficiently compressible. There have been cases of pairs of houses being extensively damaged as a result of forces being transmitted through a 50 mm wide gap between adjacent walls when for convenience the gap had been filled with expanded polystyrene. The compressibility of sheets of expanded polystyrene may be increased by removing cylinders say 75 mm in diameter on a regular grid from one side of the sheet.

DESIGN OF FOUNDATIONS AND SUPERSTRUCTURE TO TOLERATE THE EFFECTS OF SURFACE MOVEMENT
Flexible structures

As an alternative to the conventional rigid superstructure protected by a specially designed foundation, a flexible building may be provided with joints which articulate and accept the vertical movements of the ground by a readjustment of the shape of the building. A simple surface foundation able to ride over the subsidence wave is usually all that is required, which results in considerable savings in cost over conventional rigid foundations. A considerable amount of design and development work is necessary to perfect a waterproof cladding, and to produce architectural details which will perform satisfactorily throughout the life of a building subject to intermittent movement. The use of a flexible frame may affect the design of the entire building and the decision to follow this approach to subsidence design should be taken early and made known to the entire design team.

One of the best known forms of flexible construction is the CLASP (Consortium of Local Authorities' Special Programme) system which has been used for over 2000 buildings since 1956. The foundation consists of a thin flexible reinforced concrete raft with a smooth underside and no downward projections, thus facilitating relative horizontal movements between the building and the ground. The weight of the building and the coefficient of friction between it and the ground determine the amount of horizontal stress that is transmitted to the foundation. For this reason, the weight of the building is kept as low as possible by using a lightweight form of construction, and the coefficient of friction is reduced by laying the foundation on a membrane such as polyethylene. The concrete slab is made strong enough to deal with any compressive forces and is reinforced with steel fabric to withstand tension. The frame is fully pin-jointed so that any deformation will not produce additional stresses in members. Diagonal bracing in the vertical plane between columns, in some positions incorporating compressive springs, is used to resist wind loading on the structure and transmit these forces to the foundation slab: the springs resist the wind loading but yield to the building load during ground movement. The spring-braced units can go distort when ground settlement occurs while the building remains normal to the average curvature of the ground.

The roof and floors act as horizontal diaphragms to the structure and maintain its rectangular form in plan whilst being flexible enough to allow the frame of the building to follow the curvature of the ground surface. All the claddings used in the system are designed to accommodate the movement of the structural frame during subsidence and the weather joints between the vertical units will still function after movement has taken place. Neoprene gaskets are used for window glazing which help the timber window frames to accommodate deformation of the structure and to avoid breaking the glass. The partition units are fixed to the diaphragms above and below and can thus move from the vertical in one direction in sympathy with the columns; in the other direction there is adequate space inside the partition unit to allow the columns to lean in the longitudinal direction of the partition. Corner units are sufficiently flexible to permit the relative movement of panels at right angles to each other to take place. Internal doors and screens are designed to the same principles as the partitions. As the structural diaphragms ensure the building remains planar, the ceilings follow the movement of the building but remain square on plan. Staircases are articulated.

As CLASP is a component system of building, it is appropriate to large building complexes. The maximum length of continuous structure is 45 m, above which length expansion joints are introduced. Vertical subsidence of up to 1·5 m has occurred, and the ground has slid several centimetres beneath the foundations without interruption in the use of the structure and without danger to life. Remedial maintenance work following subsidence due to mining has been of a minor nature, mainly occurring in a predictable manner. It should, however, be recognized that there are other types of flexible structure than the CLASP system, e.g. timber frames.

GENERAL PRECAUTIONS APPLICABLE TO BUILDING CONSTRUCTION AND FINISHES

The following general principles applied to the design of new buildings will help to reduce the damage suffered during subsidence but it should be borne in mind that some of these principles will not apply to flexible structures.

(a) Divide large buildings into small strong units using double walls with a gap at least 50 mm wide between them, provide independent foundations for each unit.

(b) Expansion joints in structures provide convenient planes to accept strains from movements of the ground in addition to responding to thermal movements.

(c) Door and window openings should be kept to a minimum to avoid reducing the strength of walls. Door openings in structural walls should not be in line to produce a weak plane through the building.

(e) Weak blockwork for the inner skins of cavity walls should be avoided, partitions and party walls should be in strong blockwork or brickwork;

(f) Roof and floors should be securely tied to walls to produce a well braced box-type structure.

(g) Provide tensile strength at the connections between all structural members to ensure that differential displacements of the members will not produce progressive collapse.

(h) Avoid the use of arches in masonry construction.

(i) Avoid the use of corner windows, and weak projecting bays or porches.

(j) Outbuildings should be fully detached from the main building.

(k) Avoid the use of brittle finishes like plaster, use plasterboard ceilings and dry linings to walls.

(l) Provide generous falls to gutters.

(m) Paving immediately against a building should be avoided as far as possible by the introduction of planting areas or gaps. At entrances and locations where paving is essential, it should be compressible, e.g. bitumen macadam. This will avoid transmitting local concentrations of high stress on to the building, as the wave of subsidence passes.

(n) Use fences in place of garden walls.

(o) Use earth banks or dry stone walling instead of rigid retaining walls.

SPECIAL CONSIDERATIONS APPLICABLE TO
WATER-RETAINING STRUCTURES
Dams: design considerations

The special problems arising through possible or actual subsidence in the case of dams are caused essentially by:

(a) the imposition of a head of water over the floor of the reservoir and against the face of the dam;
(b) the chemical properties of the water in the reservoir which may be substantially different from those of the pre-existing natural groundwater;
(c) the imposition of the loading of the dam on the foundation.

Water seeping from a reservoir under pressure will tend to erode and enlarge underground cavities and this effect may be aggravated by aggressive qualities of the impounded water, due to the formation of new cavities by solution. Thus the risk of subsidence may be created or increased.

If the filling of a reservoir will result in a corresponding increase in pore-water pressures, the additional load from the impounded water will not be significant in terms of effective pressures. If, however, the floor of the reservoir has limited permeability or is virtually impermeable, the effective stresses will be increased by the imposition of the reservoir head with the possibility of consequent collapse of underlying cavities. The risk of subsidence of the reservoir floor may therefore be great or small depending on the prevailing ground conditions which will require careful investigation in every instance. Adverse effects due to subsidence may include:

(a) fissuring of earth dams or their foundations, possibly involving damage to filters and underdrainage or opening of joints and cracking of concrete dams;
(b) damage to ancillary works such as spillways and draw-off structures which could render the reservoir dangerous or inoperable though leaving the dam itself undamaged;
(c) indirect effects due to landslides in the reservoir slopes resulting from subsidence; these could cause surges in the reservoir threatening the safety of the dam and infilling the reservoir with debris;
(d) leakage through the dam and/or foundations.

Subsidence can reach major proportions and a further matter which should not be overlooked is the possibility that subsidence on the perimeter of a catchment might alter the extent of the catchment and consequently affect the reservoir inflow. Similarly, subsidence of the reservoir banks might allow spillage to by-pass the dam.

If subsidence occurs in such a way as to cause fissuring of the dam or its foundations, internal erosion or severe overtopping of the dam, a serious failure could result. The safety of dams is generally considered of paramount importance in cases where failure could result in the loss of human life. If subsidence were uniform and reasonably well predictable it would often be possible to allow for it and to prevent overtopping by over-building or heightening the dam, always providing that this would not cause unacceptably high lake levels before subsidence took place. However, it is rare for settlement to be uniform and more importantly in addition it is necessary to predict the magnitude and direction of surface strains which will be caused by subsidence. The practice of mining in the proximity of a dam site, but leaving a solid 'pillar' under the dam itself, is liable to result in flexure of the dam and to cause tensile strains in the adjoining ground. The prediction must therefore take into account possible future increase in the cavities by ero-sion or mining. This will enable a decision to be reached as to whether the dam can be designed to accommodate the strains, whether the subsidence must be eliminated or reduced, or whether the site is unsuitable for a dam altogether. If the possibility of subsidence is not eliminated, not only the dam but draw-off and energy dissipating works must be designed for safety before, during and after subsidence.

Prediction of subsidence is of particular importance because the opening of fissures, which might be unimportant in other structures, is potentially serious in connection with reservoirs because of the escape of water and its eroding effect. Potential sources which may give rise to erosion must be identified. These may include cavities left by rotting roots or other organic matter, joints in rock, permeable beds or strata, the holes of burrowing animals, holes formed by trees overturned in gales as well as cavities formed by the solution of minerals and the activities of man. The possibility that faults may be reactivated by mining subsidence needs special consideration in the case of dams and reservoirs. Every effort must be made to locate old workings, shafts and tunnels within the reservoir area in case they are con-nected to workings beneath the dam or downstream of it. The possibility that cavities might be created or enlarged by fires in coal seams should be investigated. Mineralogical studies must be carried out where necessary in association with forecasts of seepage water quantity in order to assess the risk of enlargement of seepage paths under chemical or physical attack by water, including the risk of extension of salt workings by solution of salt.

Dams: design methods

The principles to be followed in designing dams for areas subject to subsidence are:

(*a*) adoption of a suitable design for dam, spillway and outlet works;
(*b*) the construction of an adequate cut-off trench, grout curtain and/or impermeable upstream blanket;
(*c*) prevention of serious differential subsidence under the dam itself;
(*d*) provision for adequate drainage.

In the case of subsidence of already existing dams, remedial measures would also be directed towards achieving these aims.

The choice of design would generally give preference to rockfill over earthfill types and to earthfill over concrete types. Rockfill is less susceptible than earthfill to damage from overtopping or percolation and less subject to the effects of increased pore-pressures due to the compression of contained water which could result from subsidence. Of concrete dams, buttress types would normally be most suitable being potentially more flexible than massive gravity or arch types. To provide sufficient flexibility at joints in concrete dams, elaborate measures may have to be taken to provide water-bars, with preference being given to those which are self sealing and allow renewal if failure occurs (for instance recesses provided in the joint and filled with re-groutable sealant (Anon., 1975).

If the risk of differential settlement and/or local weaknesses cannot be eliminated, structures must be designed to bridge the discontinuities. In the case of asphalt membranes where rupture could occur, reinforcement with appropriate strength and modulus will be needed, such as polyester mesh or fabric. To provide against tensions arising from differential settlement it may be necessary to consider the use of prestressing, by cables in concrete dams and by grouting in fill dams. In fill dams, draw-off arrangements should be isolated from the dam itself so that the conduit would be unaffected by subsidence of the dam and so that the dam would be unaffected by rupture of the conduit. However, the design of offtakes in tunnels through abutments must also be adequate for the nature of the rock surround.

In cavernous ground a particularly rigorous specification must be adopted for the grout curtain and/or cut-off trench filled with impermeable material, in order to control the escape of water beneath the dam and its abutments to an acceptable amount. US Corps of Engineers' practice has been described by Moore and Couch (1970).

Measures of general application to prevent subsidence have been described elsewhere in this report. In selecting those which would be suitable for dam foundations, preference should be given to those which will provide complete support and thus avoid further settlement. It will often prove advantageous to remove particularly weak and weathered surface strata from under the dam. Filters and cores of embankment dams should be formed with cohesionless materials wherever possible, to promote a 'self-healing' effect in case of any internal damage. Fines of dispersive clay, which may be leached out to form enlarged cavities, should be avoided. The filters should form wide transition zones. Drainage provisions should be so arranged that the leakage from different sections of the dam can be isolated and measured so that the source of any increased leakage can be identified. Galleries should be provided above the line of the cut-off trench where feasible to permit inspection and remedial drilling and grouting if required. In cases where a risk of subsidence is known to exist, it is particularly important to provide an adequate installation of instruments to monitor deformations. Guidelines have been published by ICOLD (1972). The problem of establishing a datum for levels which will remain unaffected by subsidence may need special study.

Swimming baths and reservoirs

The alternatives, already discussed, of eliminating unacceptable effects of surface movement, minimizing such effects, with a relatively stiff structure or adopting a flexible construction which tolerates the effects of movement are equally applicable to water-retaining structures such as swimming baths and reservoirs. For small tanks it may be economic to provide a stiff foundation which can carry the loads in the various extremes of support conditions. In the case of large tanks, internal walls are usually essential and can be design to act as an internal grid of deep beams thereby permitting jacking pockets if provided, to be limited to peripheral locations. In a flexible design the reservoir compartments are kept small and self supporting, and connected to each other by articulated joints. The compartments may be further divided into structurally independent units with joints between adjacent sections of roof and floor. The columns or walls within these units may be designed to support the roof through pinned connections, and to transmit horizontal forces and bending moments from wind or asymmetrical vertical loading to the foundations.

For flexible designs to be successful great care must be given to the design of the joints so that they are able to accommodate the anticipated tensile, compressive and rotational ground movements. Greater initial joint widths are required than are usual in conventional reservoir designs. The net joint movements impose severe demands on water bars and joint sealants (Anon.,

1951; Lackington and Robinson, 1973). However, stiff inverted conical tanks are examples of the type of tank suitable for a design based on spanning or cantilevering over any concave or convex ground profiles and are particularly well suited to three point provision for jacking.

RETAINING WALLS AND BRIDGES
Earth-retaining structures
Earth-retaining structures should be designed in the usual manner in accordance with soil mechanics principles. (Reinforced earth and anchored structures are exceptions to this rule; see next section.) Ground subsidence affects retaining walls and bridge abutments in two ways, by withdrawal of support from beneath the structure and by imposing horizontal strains and tilts. The problems caused by withdrawal can be dealt with by assuming that the structure will span or cantilever over the subsidence area (Mautner, 1948). However, experience gained during the construction of the M1 motorway in Yorkshire has shown that application of this method to design can result in uneconomic designs, particularly if unrealistic vertical components of movement are assumed. The tensional ground strain resulting from ground subsidence does not usually require any special precautions other than limiting the size of bays or panels. The effects of compressional ground strain can be severe as the surrounding earth may be applying passive pressures to the structure. If adequate safeguards are not taken to cater for the compressional strain the resulting earth pressures may be sufficient to cause major structural damage.

The design of the retaining wall must be such as to accommodate the ground strains resulting from subsidence, and it is necessary in this context to recognize the more adverse conditions which may arise when the retaining wall is not free-standing but is integrated with a structure. It is also important to take cognizance of the possible variations in the axes of ground strains relative to the line of a retaining wall. Where such strains are principally in a direction transverse to the line of a retaining wall, differential strain and tilting can normally be accommodated by jointing a wall in panels or bays which do not usually exceed 10 m long, together with the provision of a granular sub-base layer.

Where strains may occur on an axis approximating to the line of the wall, it is necessary to provide joints at more frequent intervals and to consider the measures to be adopted to minimize frictional resistance between the back of the wall stem and the retained soil.

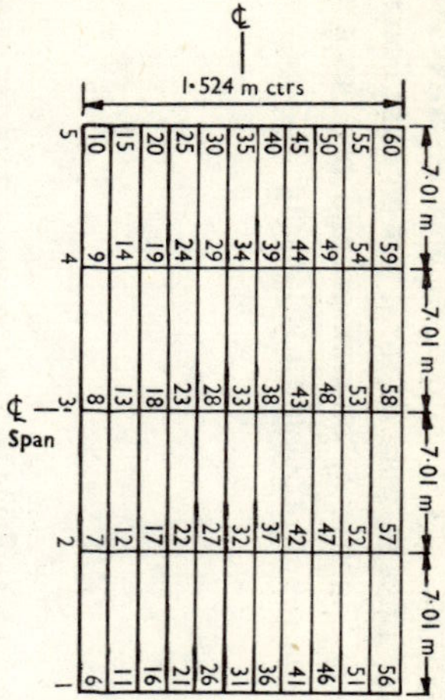

Span 29 m width 18·6 m

Loading: Type 1 HA; Type 2 20 HB; Type 3 45 HB

Note: Because of variation in loading differences in percentage overstress between types are slightly exaggerated

Type	Type and approximate section properties	Longitudinal: Maximum moment, kNm	Longitudinal: Maximum torque, kNm	Longitudinal: Maximum shear, kN	Transverse: Maximum moment, kNm	Transverse: Maximum torque, kNm	Transverse: Maximum shear, kN	Maximum moment as percentage of MDL+MLL	Calculated maximum shear stress due to torques
Type 1 Steel beam and concrete slab	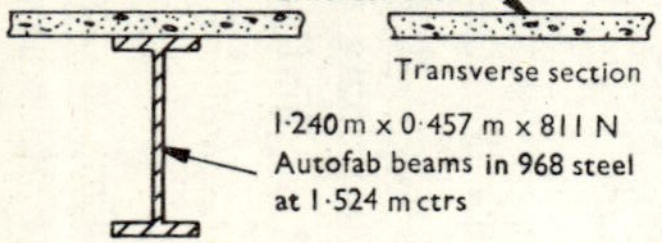 178 mm reinforced concrete deck; Transverse section; 1·240 m x 0·457 m x 811 N Autofab beams in 968 steel at 1·524 m ctrs; Longitudinal section	40·7 (Member 1-2)	16·3 (Member 26-29)	10·7 (Member 1-2)	16 (Member 1-6)	81 (Member 22-27)	17 (Member 1-6)	0·6%	Web 3·39 MN/m²; deck slab 1·0 MN/m²
Type 2 Concrete grillage and concrete slab	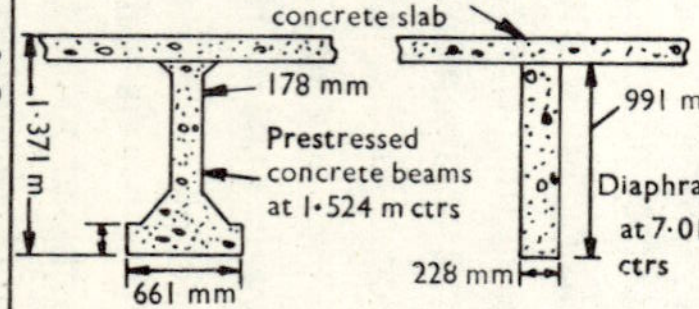 178 mm reinforced concrete slab; 178 mm; Prestressed concrete beams at 1·524 m ctrs; 1·371 m; 661 mm; 991 mm; Diaphragm at 7·01 m ctrs; 228 mm	74·5 (Member 1-2)	78 (Member 26-29)	17·3 (Member 1-2)	77·2 (Member 1-6)	111 (Member 22-27)	83 (Member 1-6)	1·0%	Web 0·94 MN/m²; deck slab 1·0 MN/m²
Type 3 Concrete box grillage	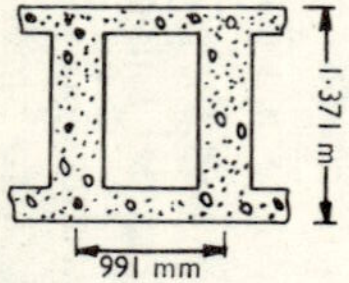 1·371 m; 991 mm; Standard PCDG box beams type 14 (A)	621·5 (Member 1-2-3-4-5)	948·5 (Member 26-29)	1717 (Member 1-2-3-4-5)	986·5 (Member 1-6)	23035 (Member 22-27)	1058·6 (Member 1-6)	43·5%	Web 0·76 MN/m²; top slab 1·9 MN/m²

Fig. 9

Special considerations relating to reinforced earth

Reinforced earth can accommodate large differential settlements and its use, in areas of ground subsidence, may prove to be a good engineering solution to a problem. The design of reinforced earth structures is usually considered in two parts:

(a) the external stability;
(b) the internal stability.

The external stability can be considered by treating the reinforced earth mass as a solid body, which corresponds to the usual conditions of stability of a gravity wall. The internal stability of the reinforced earth mass itself is considered by assuming two possible modes of failure.

(a) failure of the structure by fracture or yield of the reinforcing elements under the induced tensile forces;
(b) failure of the structure by lack of adhesion between the reinforcing elements and the earth fill.

Reinforced earth structures are susceptible to tensile strains.

Reticulated walls (Oriani, 1971) are similar to reinforced earth structures, although in this instance the structure is formed in situ by driving a large number of small diameter piles placed closely together to form a strengthened earth mass. The technique is usually limited to cuttings and as such is complementary to reinforced earth. Provided the piles forming the wall are sufficiently flexible reticulated walls can sustain all the ground strains normally caused by ground subsidence.

Special considerations affecting anchored retaining walls

Anchored walls may be susceptible to both compressive and tensile ground strains and their use in areas of major subsidence should be considered with care.

Design of bridge substructures

The design principles outlined in the earlier sections of this chapter are applicable to bridge abutments. Precautions should be taken to cater for any additional vertical and horizontal loads caused by ground subsidence which may be applied through the bearings, and to cater for any horizontal movements in the deck. Jacking pockets, or other provision to facilitate jacking, should be provided to help maintain or restore the deck to a satisfactory level either while the subsidence is taking place or when it is finished. Care should be taken to ensure that the subsidence does not disrupt the abutment drainage.

Design of bridge superstructures

Subsidence can produce three-dimensional movements which if transmitted to the supports of a bridge will cause relative displacements in all directions and subject the bridge to tensile and compressive strains. The bridge could be designed to resist the ground movements but it is usually more economic and satisfactory to articulate it so that the subsidence has little effect on the distribution of stresses. The differential movements and stresses to which a bridge may be subjected during subsidence are summarized below, together with typical values:

(a) differential longitudinal horizontal displacement in which the bridge suffers either compression or tension along its length caused by the supports moving together or apart ($\pm$ 225 mm);

(b) differential transverse displacement in which the supports move sideways parallel to each other ($\pm$ 150 mm);

(c) differential vertical displacement in which the supports settle differentially ($\pm$ 1·0 m);

(d) differential longitudinal tilt in which the supports tilt towards or away from each other along the axis of the bridge ($\pm$ 1 in 80);

(e) differential transverse tilt in which the supports tilt at right angles to each other ($\pm$ 1 in 150);

(f) differential rotation in plan, in which the supports rotate about their own vertical axis. ($\pm 0° 20'$).

In certain cases the effects may be only two-dimensional, for example if a seam is mined by modern long wall mining along the line of the bridge, the movements (b) and (e) may be negligible. The structure would then only need to be two-dimensionally statically determinate. For this situation to arise with a skew bridge, the mining would have to be at right angles to the piers.

In general the effects are three-dimensional and it is necessary either to make the bridge torsionally weak so as to withstand the three-dimensional effects or to make the bridge fully three-dimensionally statically determinate. The effect of twisting on short span bridge decks (30 m) has been reported by Sims and Bridle (1966). In this study a simply supported bridge superstructure span 30 m and width 18 m was subjected to a twist such that the slope of one support relative to the other was 1 in 80. It was found that this could be withstood by the low torsion decks but not by the high torsion deck (Fig. 9). The three decks were as follows: longitudinal steel beams and reinforced in situ concrete slab (low torsion); prestressed concrete I beams with reinforced in situ concrete slab (preferably without internal diaphragms) (low torsion); prestressed concrete boxbeams transverse prestressed with reinforced in situ concrete slab (high torsion).

Particular care needs to be applied to the design of bearing and expansion joints. They must be able to take all movements including the tilting, twisting and transverse movements. Special consideration should also be given to the detailing of parapets which may be affected at joints. Wherever possible, an adequate allowance should be made for movements. However, local damage to parapets and also carriageway surfacing may need to be accepted during the passage of mining.

MONITORING THE EFFECTS OF SUBSIDENCE ON STRUCTURES
General
Monitoring will involve a series of observations of a structure which may commence before or after the onset of subsidence. Observations should include the recording and measurement of cracks, precise levelling of the structure, and may include the measurement of horizontal movement in the structure or soil. Reference should be made to Burland and Wroth (1975) who deal comprehensively with all aspects of cracking and differential settlement of buildings. They also define settlement, heave, differential or relative settlement, rotation, tilt, relative rotation, angular strain, relative deflexion and deflexion ratio.

Inspection of structures
At each observational inspection of the structure a record should be made of the positions, approximate inclinations and first date of observations of any crack. Observation by the naked eye will generally be sufficient up to a height of 4 m, supplemented by binoculars for heights up to 12 m. Above this height special access arrangements will normally have to be made, possibly in the form of a mobile hoist (Cheney, 1974). The recordings of cracks can be made conveniently on copies of constructional drawings.

The width of cracks detected by the above methods will vary with the building material, but generally those greater than 0·2 mm will be seen. An inspection revealing no cracks or no new cracks should be treated as a positive monitoring operation, the date being carefully recorded with those of other observations.

Measurement of cracks
The use of glass tell-tales across cracks is deprecated. When embedded in

mortar, they often fail to indicate due to anchorage slip and, when attached by epoxy resin, are still subject to vandalism or other false indication. Simple measurement of crack width is best made using a steel rule, preferably chrome faced for legibility and marked in full millimetres on both edges of one face only. The position of measurement should be permanently marked on the structure, one method being by pencil mark on a small area of emulsion or weatherproof-type paint. As an alternative to a steel rule, a pocket magnifier equipped with glass graticule having etched scales graduated to 0·1 mm may be employed (Cheney, 1974). The magnifier is used with the glass scale in contact with the brickwork. In order to record in more detail the relative movement across a crack in the plane of the surface, two measurements may be made at right angles to each other.

One method involves setting into the wall three brass round-headed head screws with their heads protruding about 5 mm; No. 6 × 30 mm is a suitable size. Normal fibre plug fixing may be employed providing that the plug face is recessed about 10 mm and the void so formed filled with epoxy resin before setting the screw. This will ensure lateral stability of the screw in all climates. The screws are positioned to form the corners of a right-angled triangle of side about 75 mm and which straddles the crack in one of the configurations shown in Fig. 10. In configurations (a) and (b) the horizontal (h) and vertical (v) components of crack movement may be measured directly with a 150 mm vernier caliper and the resultant vector of movement calculated if required. Where, due to the direction of the crack, a configuration such as (c) or (d) is utilized, the resultant vector of movement may be similarly calculated or if required the horizontal and the vertical components of the crack movement obtained. The three-screw method is subject to second order complications as the triangle distorts but it will not usually be considered necessary to correct for these. Other methods of crack measurement involve the use of dial gauge systems such as the Demec gauge (Building Research Advisory Service, 1971).

Levelling

Levelling will involve the installation of purpose-made levelling stations and the use of a precision level. Cheney (1973) covers in detail all aspects of the levelling of buildings for monitoring purposes. This reference deals with equipment and techniques for those who wish to undertake precision levelling of structures. It has a description of and gives installation details for the BRS levelling station as shown in Fig. 11. Names of current commercial suppliers can be obtained from BRS.

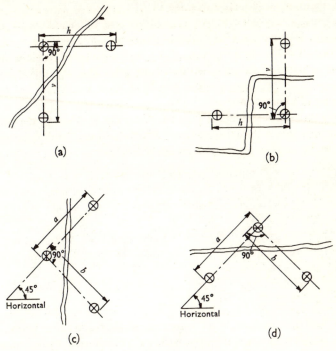

Fig. 10. Configurations of screws for crack measurement

Datum points for levelling

The requirements for a datum are covered in Cheney's (1973) publication and this should be referred to. It is to be noted that a datum may consist of a purpose-installed rod and tube system, a deep reference system such as derived from the extensometer system described by Smith and Burland (1976), or a BRS levelling station fixed to a stable building in the vicinity. In all cases it is advisable to employ two or even three datum points as a check on each other. It should also be noted that as it is differential settlement which will damage a building any instability of the datum system will not effect the accuracy of differential settlement calculations made from the levelling. Therefore although it is desirable in all cases to establish as stable a datum as possible, even with a known instability it is still worthwhile using the most precise levelling techniques. In the case of levelling associated with mining subsidence, a stable datum can be established only in consultation with the National Coal Board.

Measurement of horizontal movement

The measurement of horizontal movements in structures and the soil may be appropriate in some instances. Several methods are described by Burland and Moore (1973). In particular their micrometer rod system may be adapted for use directly on structures using BRS levelling stations.

Frequency of observations

The frequency of observations may vary considerably during the course of the study of a given structure and from structure to structure. A programme should be followed covering at least the following:

(a) Stages of any continuous works liable to affect the structure, e.g. at regular intervals of depth during excavation, at each floor of a construction, at intervals of advance of mining beneath the area, at stages of underpinning work.

(b) Before and after relatively rapid engineering load changes, e.g. jacking of struts, removal of strutting, tensioning of ground anchors, casting of large masses of concrete.

(c) Long-term effects due to works as in (a) and (b) above, and those due to vegetation changes in clay soils etc.

As a budgetary indication a programme might resolve into taking observations between twelve and fifty times a year during active mining or engineering works followed by two a year for one or two years with the period lengthening to a year thereafter until it is apparent that no significant movements are taking place.

PROVISION IN NEW STRUCTURES FOR FUTURE RE-LEVELLING BY JACKING AND IMPLEMENTATION OF JACKING OPERATIONS

General guidance on provision for future jacking

As a general guide the cost of making provision in a structure for future re-levelling is likely to be least when the jacking points are numerous and grouped to match the load of the structure. However, more points of access to jacking positions will be required. This is often acceptable when the likelihood of jacking is remote, for example, where an estate of houses needs to be protected although only a small, but unpredictable, number of them are likely to need re-levelling. Increasing the number of jacking positions will increase the control problem during a lifting operation but this may be readily overcome.

When designing for jacking, the optimum distribution of jacks must be

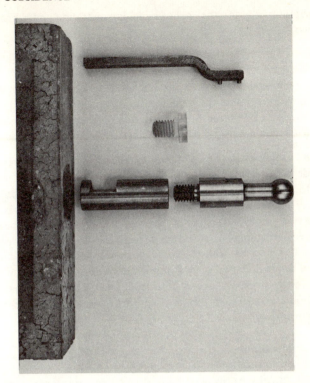

Fig. 11. Components of BRS levelling station

considered in relation both to the structure that is to be maintained level and the foundations which will support them, as well as the loads expected on the jacks, the rate at which these loads can be expected to change as the structure distorts and the pattern, rate and magnitude of vertical and horizontal ground movement which may be expected. Occasionally, it may be desirable to install a permanent and possibly fully automatic jacking system beneath a structure in order to isolate it from the effects of ground movement. In such a case the engineer must liaise closely with the designers, installers and operators of the equipment. Permanent access to the equipment will be required and a basement or subfloor voids will almost certainly be required. In most cases, however, a building or structure may be made jackable with relatively minor modifications. Many buildings in which provision for jacking has been made have not distorted sufficiently to warrant later lifting and thus the cost of special provisions made at the time of construction should be kept to a minimum even if this increases the cost of the jacking and repair work, should it be required later.

It is helpful to classify the cost of a jacking operation under four headings as follows:

(a) cutting away, removing covers, etc. to provide access to the work and making good or replacing on completion;
(b) temporary and permanent modification and/or strengthening of the structure to facilitate the jacking;
(c) installing and operating the jacking and level control equipment;
(d) repair of damage to the structure caused by the subsidence.

In the case of domestic and commercial buildings it will usually be desirable to provide a suspended ground floor, as this will minimize disturbance during lifting. Similarly ground floor areas in an industrial building supporting closely spaced equipment which is not easily moved, such as boiler house floors, may be suspended so that in the event of jacking there will be no need to disconnect the equipment. A suspended ground floor will also provide additional stiffness to the substructure which will help to resist dimensional changes of the building on plan. Foundation sizes must be chosen to accommodate lateral movements due to subsidence and the transfer of loads from the initial support positions to the jacks and to the temporary and permanent packing positions.

Design considerations
A building or structure which is required to be lifted must do so without jamming or sticking to the existing base. This requires considerable care in detailing and joints and gaps should be larger than is generally the case for expansion joints. Joints between sub-base and structure should be designed to become wider as the lift proceeds. This requires that, where practicable, upright joints should be angled and not parallel to the direction of lift. If a structure is likely to become tilted and is surrounded by walls, such as a building with a basement or an isolated section of a larger structure, then the gaps between the structure and the fixed walls will close as the tilt is corrected and they must be large enough to accommodate this. A minimum gap of 50 mm at maximum distortion or after lifting or levelling individual sections is recommended in all vertical or inclined joints. The minimum tolerance in gaps between adjoining structures which may be tilted differentially should be increased by 10 mm per metre of joint.

The designer must be absolutely certain that all joints are clear. Inspection during and on completion of construction should be specified and its execution should be checked. The detailing should provide cover strips to prevent joints from becoming filled with dirt or debris. Horizontal joints

below the base of the structure must also be carefully considered. Suction
and adhesion forces between the structure and sub-base can be large and if
widely distributed they may damage the base of the structure. Some
materials may tend to bond together under pressure sustained for long
periods and such materials should be confined to low pressure areas and
used in multiple layers. For high pressures use should be made of steel plates
resting on one another. Alternatively, dry joints in masonry or joints of
weak lime mortar may be provided. A raft laid on a sand bed will always lift
off easily, as would a joint of hardboard on 'no fines' concrete.

In most cases the jacking operations are undertaken long after the design
of the installation. The jacking equipment available may have changed and
the designer is advised to allow for pockets which will accommodate jacks
of alternative designs, together with adequate working access to place the
jacks. Jacks weighing more than about 25 kg are difficult to handle in confined spaces and very heavy jacks will need provision for winching or
moving the equipment into position with trolleys. A minimum clearance of
30 mm in the height of jacking pockets plus the expected construction
tolerance is recommended. If only small lifts are anticipated shallow pockets
for flat jacks may suffice but deeper pockets for longer stroke cylinder jacks
will generally be more suitable.

When choosing jacking equipment allow for the equipment to work at not
more than two thirds the specified working load of the jacks. If the building
is very rigid one half working load is more appropriate but for a highly articulated and determinate structure, or a three point system, three quarters
of working load may be used. Under very heavy buildings 'follower' jacks
on separate circuits working at 80% of working capacity may be included in
the system. For heavy buildings hydraulic jacks with power-operated pumps
will be required and high-pressure hydraulic mains will often be needed
linking the jacks, preferably run on a direct route between them. Hydraulic
jacks are subject to collapse if the oil seal fails or the oil supply line breaks.
Oil line failures can be guarded against by non-return valves but alternative
means musts be provided for releasing the oil pressure when retracting the
jacks.

The system must be designed to be 'fail safe' and where heavy loads are
involved duplicated circuits may be required. Threaded collars on the rams
are recommended particularly when jacks are used singly. Screw jacks with
a high mechanical advantage which will lock in position under load and will
only lift or retract when operated should be considered for light structures.
The system chosen to control a jacking installation will depend on circumstances. If a direct lift is required all the jacks will need to lift at the

same rate provided that the foundation supporting the jacks is stable and will not settle differentially under the jacking. If a building is tilted, the various jacks will need to move at different rates in order to correct the tilt. This will also be the case if the subsoil is weak or disturbed and can be expected to sink differentially during the jacking.

The simplest system to control is a three point system, in which case a plumb line or optical plummet may be used as an alternative to conventional levelling. The jacks may be operated one support at a time to bring the structure back to plumb. If the foundation is certain to remain undistorted during jacking then premeasured lifts calculated after surveying the structure will often be satisfactory. With a large number of jacks distributed singly or in numerous groups a liquid level system will be effective; this may be a moving liquid level system designed to guide each group of jacks or a fixed liquid level system designed to give a datum at each jacking position (Charge, 1972).

It is important to monitor and control a lifting operation using a single datum point which will not be disturbed by the jacking, in order to eliminate the effect of possible differential foundation movement. It will sometimes be more convenient to monitor the levels on the first floor of a building or the ceiling above ground floor level, rather than the ground floor. Lowering a structure is generally a more difficult operation than raising it, as most equipment is designed to lift easily but it is more difficult to release oil under close control from a jack under pressure. It also requires more care with the packings when lowering as these must be progressively removed before they become jammed. Screw jacks avoid these difficulties Beurak *et al.* (1966).

If fixings or pockets for the jacks are to be kept small, that is, less than 500 mm deep, then short-stroke jacks will be required with an effective lift of the order of 200 mm. Thus jacking will generally proceed in repeated sequences of operations which include lifting the building, retracting the jacks and resetting them with packings above ready for the next lift. Provision must be made for placing and tightening packings at separate but adjoining positions if jacks are used singly. If groups of jacks are used beneath a rigid support, they may be retracted, packed and tightened back in position in sequence leaving the remaining jacks to carry the additional load during the packing period, provided that the release of one jack would not cause the remaining jacks to become overloaded.

To guard against hydraulic failure during lifting either use separate packings raised in step with the lift or jacks with screwed collars which are tightened down at frequent intervals. At the end of a relevelling operation the jacks are most economically replaced with concrete blocks which must be pinned up to the work above. Also a gap will have opened above the original support and this must be sealed or pinned up.

Implementing a jacking operation

The first stage of implementing a jacking operation will be to take levels and prepare a structural survey. This may include regular monitoring of the movement as it approaches the critical amount. The total relevelling and lifting procedure should have been thought through and recorded at the design stage. Such records should be carefully studied before implementation. In important structures, or where subsidence is a near certainty, it may have been desirable to carry out a trial lifting operation at the time of construction. However, in most cases the jacking operation may not occur until many years have passed, if at all, and faults or omission in the design are then very expensive to correct and may to a large extent invalidate the provisions that were taken. Particular attention must be given to checking on site to see that voids and gaps were properly constructed, and that all pieces of formwork or other objects that could cause the structure to jam are removed, that no dirt or debris has fallen into the gaps, and that no cracks that will need to close have been made good.

Access to the jacking positions will need to be opened. For example, this may involve trenching out around the building and removing floor panels. The next stage of preparation will be to install the extra members or to modify the building as required. Thus, for example, the bottoms of portal frames may need to be tied before bases are released or tall structures may need to be guyed. The lifting equipment, hydraulic system, monitoring and level control equipment should then be installed and tested up to full working pressure. Further packings should be to hand and adequate tested reserve equipment should be available.

The sequence of operations for lifting will be predetermined. Monitoring and control of levels during lifting or lowering should concentrate on the structure being moved and should not relate solely to the sub-base which may move irregularly under pressure, particularly if the ground has been disturbed by subsidence. Lifting in predetermined amounts related to the jack foundations and calculated from an original level survey will be effective only if the jack foundations are adequate and firm.

A structure is likely to be at its most vulnerable when being lifted and attention must be given to the risk of high winds. It may be that the operation will need to be stopped, packings placed, screw collars tightened down and guys tightened in such circumstances. Care must also be taken to ensure that the structure is left secure during breaks and overnight. The equipment installed must provide for this contingency.

It cannot be emphasized too strongly that the safety of both men and structure is a prime consideration in such an operation. The work must be kept tidy to avoid accidents and to facilitate quick action should some part of the equipment fail. Adequate packing should always be to hand at every support point. Careful detailing and adequate preparation for such operations will be repaid during the lifting. On completion of relevelling, the jacking equipment will be removed in sequence and replaced with permanent packings. On completion of reinstatement a report on the work should be deposited with the building records, with particular attention paid to the possibility of further jacking.

RELEVELLING TILTED AND DISTORTED STRUCTURES IN WHICH NO PROVISION FOR JACKING HAS BEEN MADE
General
It is not widely known that methods of relevelling tilted or distorted structures are well developed and very economical compared with total or partial demolition and reconstruction to similar designs and specifications.

Any structure can be lifted provided that it is adequately strengthened and most structures can be relevelled without additional strengthening of the superstructure provided that a sufficient number of jacks are used. Experience indicates that, for brick or masonry buildings, the most effective and economical solution will be to construct at the base of the structure to be lifted a light framework which will provide support between more widely spaced jacks and which will tie the building together at the base; some bracing at higher level may also be required but normally need not be extensive. In framed structures the jacks will be best grouped around the columns which may also need to be tied.

When dealing with existing structures the engineer may be required either to devise a scheme to protect the building against expected future subsidence, for example approaching disturbance from deep mining, or to relevel and repair a building which has been affected by subsidence.

Provision for future relevelling by jacking of existing structures
The expected disturbance must be studied and the optimum solution arrived at by first determining the standard of protection required and then the cost effectiveness of alternative protective measures. If it is anticipated that the structure will become permanently tilted more than say 1:100 or that the

angular distortion within the structure will exceed 1:150 then provision for jacking is required.

The base of the structure should be strengthened by a continuous reinforced concrete frame or by an equivalent permanent tying and stiffening structure designed to resist tension and compression while the ground slides beneath it (Pryke, 1954).

The structure may be jacked up from below the frame after subsidence is complete to correct the tilt, or during the subsidence cycle to minimize damage and disturbance. It may be desirable to cut joints in the structure at intervals and to provide temporary or permanent stiffening frames at these joints. It is particularly important to separate parts of a building of very different mass and stiffness, for example the tower and nave of a church. Methods of inserting structural frames beneath existing buildings are well developed and reference should be made to texts and papers on underpinning (Burland and Moore, 1973; Pryke, 1954, 1974; White, 1962; Tomlinson, 1975; Charge, 1972).

The optimum method of strengthening and modifying an existing structure to protect it against future subsidence may be different from that which would have been chosen if the modifications had been built in during construction, but the 'new building' solution is the best starting point for a design study.

Repairing damage to existing structures by jacking

The cost of structural strengthening and renovating low-rise domestic buildings which are tilted so severely that they are effectively uninhabitable is generally about half the cost of rebuilding to a similar specification. Greater savings may be expected with high buildings and framed structures.

Before executing remedial works on a damaged building it is essential to investigate the history of the damage, to identify the causes, and to consider the possibility of further movement. The building must be carefully surveyed in order to identify the amount and pattern of distortion, to record and understand the construction, and to judge the quality of the materials. The foundations and subsoil must also be investigated and information gathered about old and current underground workings and the geology of the area (McFarlane and Tomlinson, 1974). When analysing the damage it should be noted that differential vertical movements at foundation level generate rotations and most parts of the structure above foundation level will move both horizontally and vertically.

It is important to distinguish between damage due to foundation failure and settlement and distortion due to ground subsidence. If deep mining is sidence has occurred over a protracted period and extensive making good

STRUCTURES header

suspected then continued careful monitoring of movement in relation to the mining programme will help to establish whether, and the extent to which, mining is a contributory factor. If the foundations are failing then they will need to be strengthened.

It is usually impracticable to straighten out a building by jacking if subsidence has occurred over a protracted period and extensive making good and rebonding has been carried out in the past. However, any general tilt which has developed may be corrected economically together with any recent distortion that has not been repaired. Many cost effective repairs to tilted low-rise domestic and commercial structures with load-bearing brick or blockwork walls have been carried out using a light temporary framework inserted just above the foundations to tie and support the building, which is then supported in the relevelled position by pinning up from the existing foundations. Structures suffering severe curvature may best be straightened using individually controlled closely spaced jacks without a supporting frame.

A permanent reinforced concrete frame may be preferable under masonry buildings or if a large permanent lift is required to raise a building to a new level, for example to correct problems of flooding. It may even be necessary to lower a section of a building to match the remainder. It will usually be necessary to tie the building at the base when relevelling by jacking and sometimes positive horizontal thrust designed to close up cracks at base level can be effectively applied.

REFERENCES

Anon. (1951). Large reservoir designed for mining subsidence. *Concr. Constr. Engng* **46**, 353–358.

Anon. (1967). Prestressed concrete reservoirs designed for a mining subsidence area. *Building,* March, 155–156.

Anon. (1975). Soft rock makes hard problems on Panama's biggest project. *Engng News Rec.,* 29 May.

Bell, S. E. (1977). *Successful design for mining subsidence.* Conference on large ground movements and structures Cardiff.

Beurak, G., Hurst, G. & Owen, F. (1966). *Some observation on mining subsidence on large schools.* Thesis, University of Sheffield.

Bjerrum, L. (1963). Discussion. *Proc. Fifth Int. Conf. Soil Mech. Fdn Engng* **II**, 135–137.

Building Research Establishment. (1971). *Devices for detecting changes in the width of cracks in buildings.* Report by Building Research Advisory Service, publication No. TIL5. Watford: Building Research Establishment.

Burland, J. B. & Moore, J. F. A. (1973). The measurement of ground displacement around deep excavations. *Proc. Symp. Field Instrumentation,* 70–84. London: British Geotechnical Society. Republished as BRE Current Paper No. CP 26/73.

Burland, J. B. & Wroth, C. P. (1975). *Settlement of buildings and associated damage.* BRE Current Paper No. CP33/75. Watford: Building Research Establishment.

Charge, J. (1972) Raising of Old Wellington Inn and Sinclair Oyster Bar. *Struct. Engr* **50,** No. 12, 483–494.

Cheney, J. E. (1973). Techniques and equipment using the surveyor's level for accurate measurements of building movement. *Proc. Symp. Field Instrumentation,* 85–99. London: British Geotechnical Society. Republished as BRE Current Paper No. CP26/73. Watford: Building Research Establishment.

Cheney, J. E. (1974). Discussion. *Proc. Conf. Settlement of Structures, Cambridge,* 768–770. London: Pentech Press.

Ciesielski, R. & Czosnowski, M. (1977). *The construction of a passenger chair lift in a mining subsidence area and its protection.* Conference on large ground movements and structures Cardiff.

ICOLD (1972). *Report of Committee on Observation of Dams and Models.* Bulletin 23, July.

Institution of Structural Engineers (1977). Structure–soil interaction—report of *ad hoc* committee.

Institution of Structural Engineers. (1975). *Technical report: design and construction of deep basements.* London: Institution of Structural Engineers.

Lackington, D. W. & Robinson, B. (1973). Articulated reservoir in mining subsidence areas. *Jnl Inst. Water Engrs* **27,** No. 4, 197–215.

McFarlane, I. H. & Tomlinson, M. J. (1974). Site investigations for structural foundations. *Struct. Engr* **52,** No. 2, 57–65.

Mautner, K. W. (1948). Structures in areas of mining subsidence. *Struct. Engr* **26,** January, 35–69.

Moore, E. C. & Couch, F. B. (1970). Earth embankment design and foundation treatment on a highly solutionized limestone. *Tenth Int. Congr. Large Dams* **II,** R20, Question 37.

National Coal Board. (1975). *Subsidence engineers' handbook.* London: National Coal Board, Mining Department.

Oriani, M. (1971). Road embankment stabilization by reticulated walls. *Civ. Engng Pub. Wks. Rev.,* September.

Polshin, D. E. & Tokar, R. A. (1957). Maximum allowable non-uniform settlement of dwellings. *Proc. Fourth Int. Conf. Soil Mech. Fdn Engng* **I,** 402–405.

Pryke, J. F. S. (1954). Eliminating the effects of subsidence. *Colliery Engng* **31,** No. 37, 501–507.

Pryke, J. F. S. (1974). Differential foundation movement of domestic buildings in South East England—distribution, investigation, causes and remedies. *Proc. Conf. Settlement of Structures, Cambridge,* 403–419. London: Pentech Press.

Skempton, A. W. & Macdonald, G. H. (1956). The allowable settlement of buildings. *Proc. Instn Civ. Engrs* **5,** Part III, 727.

Smith, P. D. K. & Burland, J. B. (1976). Performance of a high precision multi-point borehole extensometer in soft rock. *Can. Geotech. Jnl* **13,** No. 2, May, 172–176. Republished as BRE Current Pater No. CP41/76.

Starzewski, K. (1974). Discussion. *Proc. Conf. Settlement of Structures, Cambridge,* 808–810. London: Pentech Press.

Tomlinson, M. J. (1975). Shoring and underpinning. In *Foundation design and construction, 3rd edition,* Ch. 12, pp. 722–746. London: Pitman Publishing Co. Ltd.

Wasilkowski, Inż. F. (1951). *Complete protection of structure against damage due to mining subsidence.* Translated from the Polish and reprinted by Cement & Concrete Association. Translation No. 55.

White, E. E. (1962). Underpinning. In *Foundation engineering,* Ch. 9, pp. 826–893. Tokyo: McGraw Hill Book Co. Inc.

Chapter 5
Communications

GENERAL
Modern means of communications include highways, railways, tunnels, canals and airports. All of these are, in general, affected by ground subsidence, the permissible degrees of which vary, but in each case, large ground movements are rarely acceptable. It is, therefore, necessary to adopt preventive measures where practicable in order to obviate the high cost of reinstatement.

EARTHWORKS
The execution of civil engineering works related to the various types of communications generally entails earthworks involving large and expensive items of plant and equipment which, in subsidence areas, introduce two conflicting aims (for economical use of the plant). Maximum speed of operation is required generally but safety requirements call for great care and in many cases a slower operation. In the case of high speed vehicles such as motor scrapers and tipper trucks, the possibility of wheels sinking into an unexpected cavity entails a danger to life and limb and damage to the machine. It is therefore essential that operators of plant and equipment are subject to speed restrictions in such conditions.

Where shallow mine workings of the pillar and stall type exist, it is usually possible to receive some warning when the excavation is nearing the workings as the ground tends to move under the wheels and all site staff should be trained to keep constant watch for irregular ground movements. Such movements are particularly noticeable under the wheels of laden tipper trucks and autograders.

Where the condition of workings is not known such as in the case of bell pits, a technique has been developed wherein a motor scraper is carefully used to excavate a deep trench across the site of the workings, thus enabling them to be observed in 'cross-section'.

HIGHWAYS

The highway is deemed to consist of carriageways, footways, verges, kerbs, surface water drainage, fences, cycle tracks and street furniture. It must also be noted that the highway generally contains other services such as water mains, electricity and telephone cables, sewers and gas mains, which have to be taken into account.

All the foregoing are affected by mining subsidence and examples of the effects are as follows:

(a) distortion in horizontal and vertical alignment;
(b) severe irregularity caused by uneven settlement of carriageways associated with faulting or shallow workings;
(c) fracturing of carriageway leading to deterioration of subgrade and carriageway pavement necessitating reconstruction;
(d) undulations on carriageway surface;
(e) fissures in subgrade causing carriageway deterioration;
(f) damage and displacement of flagging and footpaths;
(g) damage and displacement of kerbs and channels;
(h) damage and displacement of boundary fences;
(i) damage and displacement together with alterations of gradients of the surface water drainage system and consequential flooding of the highway or adjoining land;
(j) flooding of the highway solely due to settlement of carriageway surface;
(k) damage to the highway following bursting water or gas mains caused by pulling of joints or fracturing of pipes, also failure of underground cables.

The most common types of subsidence damage are undulations and distortions of carriageways with cracks and similar damage to footways, caused by compressive and tensile stresses in the subgrade. In the case of new highways many of the effects of subsidence may be obviated or minimized by taking special precautionary measures at the design stage as the effects of subsidence due to modern methods of extraction are predictable.

Design of new highways

In the design of new highways, the effects of shallow workings play a dominant part and it is essential that, at or before the design stage, information should be obtained concerning their location. Apart from shallow mining operations, it is necessary to contact the National Coal Board to ascertain what past deep mining has occurred within the area of the new highway, and to ascertain their long-term planning proposals.

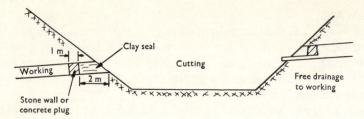

Fig. 12

At the design stage, allowance should be made for extra land to be acquired to cater for later adjustment of embankment level where subsidence may occur during or after construction. Difficulties may also occur due to subsidence affecting ground levels between the time of the initial survey and the period of construction, and monitoring should be carried out at frequent intervals to verify the accuracy of the permanent and temporary bench marks. This is extremely important if information from earlier aerial surveys is used for design purposes.

When the route has been decided, it is essential that long-term cooperation with the National Coal Board should continue, in order that the appropriate method of mining can be employed to minimize the problems which will arise if mining continues during the construction stage and immediately after. One of the most important matters to be considered is the effects of subsidence on the drainage system and the actual alignment and curvature of the motorway system, and it is by collaboration with the Coal Board that problems can be overcome.

CUTTINGS AND EMBANKMENTS

The effects of shallow mine workings within cuttings require careful consideration of the following problems:

(a) The possibility of roof collapse resulting from exposure of the working to the atmosphere. Where pillar and stall methods have been adopted, the roadways may stand for many years if air is excluded from the workings.

(b) Subsequent failure of the cutting slope as a result of the problems mentioned in (a).

(c) Preventing the ingress of air where there is a risk of spontaneous combustion.

(d) Preventing the ingress of water to workings at lower levels in cuttings and maintaining free drainage in the workings at higher levels.

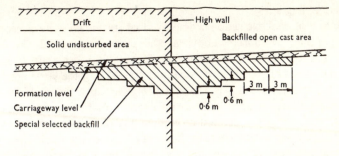

Fig. 13

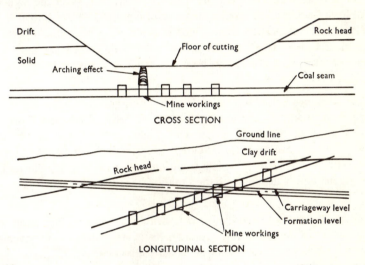

Fig. 14

Openings in the cuttings must be permanently sealed by the application and use of stone walling or the construction of a concrete barrier between the rough site shuttering (Fig. 12). The barrier should be solid, approximately 1 m in thickness, and subsequently covered with a plug of clay prior to the topsoiling of the side slopes of the cuttings. In workings at higher levels dipping to cuttings, a drainage outlet should be installed at floor level and connected to the main carriageway drainage system. The water percolating from the workings may contain sulphates or other deleterious agents, and requires investigation, but in any case, it is a wide precaution to use sulphate-resisting cement in concrete barrier walls or in mortar for stone walls.

Opencast mining sites

This method of coal extraction is employed where shown to be of economic advantage. The overburden above a coal seam is extracted, the coal is worked and the overburden replaced. It has been generally estimated, although no research has been carried out, that 90% of the settlement of the overburden takes place within the first twelve months, and residual settlement can continue for up to ten years. Where a motorway, or main roads, cross an opencast site, a method of compacting the area has been evolved, whereby the carriageway crosses the high wall or the area where the extraction of the coal has ceased. Experience has shown that a generally satisfactory method of treatment is to excavate the area in the vicinity of the high wall and to form benches 3 m wide, with a step of 6 m in height, as shown in Fig. 13. The area of excavation is then filled with a class A fill and compacted to a high degree, with less than 5% of air voids. This will reduce differential settlement after the highway has been constructed and open to traffic.

Shallow workings below formation level in cuttings

Where these workings are encountered at or below formation level, it is ncessary to take measures to prevent subsequent collapse of the road following progressive cavitation. Sudden and irregular subsidence of formation can be experienced. A typical situation is illustrated in Fig. 14 and the following method of dealing with the situation has proved satisfactory:

(a) Where the roofs of workings occur within 5 m below formation level, the whole of the underlying ground is excavated down to the pavements of the workings. If roof collapse has already occurred and the workings are found to be filled with debris, the depth of excavation may be reduced as the passage of the heavy compaction plant will compact the debris, although care must be taken to ensure that it conforms to the standard specified for an embankment. A stone or concrete wall with a clay plug must be constructed across any gallery openings at the low end of the seam to prevent water draining into the workings. The same may have to be done on the high side if there is evidence of the presence of water in the workings, or if levels permit, it may be led into the highway surface water drainage system.

(b) Where workings occur at distances greater than 5 m below formation level to roof of working, a suitable reinforced concrete slab may be incorporated in the construction of the carriageway immediately below the base, replacing sub-base material. A typical slab would be 200 mm

thick, extend 1·2 m beyond the edge of the carriageway and be reinforced with top and bottom layers; each layer comprises 5 kg of reinforcement per square metre of slab. Alternatively, the workings may be
grouted with a p.f.a. cement grout as described in Chapter 3 and under
certain conditions, this solution may be more economical.

Various combinations of these methods may be found advantageous depending on the particular site conditions, and it is a matter for the engineer to
decide on the most satisfactory and economical solution.

EMBANKMENTS
Shallow mineworkings below embankments
Where workings are known to exist below the site of an embankment within
5 m from natural ground level to roof of workings, the whole should be
excavated to the floor of the workings and backfilled as described for cuttings. As mentioned before, a check must be made for the presence of water
in the workings and, if found, it must be dealt with to prevent adverse effects
which might result from interrupted flow. Where the workings exist below a
depth of 5 m a careful analysis should be made of ground stability relating
rock cover to thickness of seam. Generally, the overlying drift and solid
varies considerably and a detailed examination of the site investigation information must be made and, where desirable, a trial trench excavated to
ascertain the nature of the soils. If the drift is of a stiff cohesive nature, it will
possess a certain amount of 'beam' strength which will assist in promoting a
bridging effect over what might otherwise be a sudden and serious oc-
currence of local subsidence. In granular soils, any collapse of the rock
strata above the mine workings will promote an immediate settlement approaching the angle of repose. Careful assessment of the state of granular
soils must be made and where there is any doubt, advice should be sought
and consideration given to preventive measures being taken, such as
grouting of the workings and the above depths may have to be materially increased. Where bell pits are found, their condition can be determined by
employing a scraper to excavate a deep trench to the bottom level of the pits
in order to expose them in 'cross-section'. Where an embankment is to be
constructed over them, it is likely that the passage of heavy earthmoving
and compaction plant may well make any further work on them un-
necessary; however their condition must be confirmed before arriving at a
decision.

Drainage
Surface water drainage systems should be constructed with pipes having

flexible joints with sufficient tolerances to accommodate movement caused by mining subsidence. It is advisable to use short pipes of the order of 1 m in length, in order to introduce flexibility into the pipe line. All surface water drainage systems should be laid on a bed of granular material and at manholes on the pipeline, there should be two flexible joints immediately outside of the manhole to form a movement joint between the relatively rigid manhole and the pipeline.

Where practicable, the drainage system gradient should be increased to ensure a continuous flow after subsidence has occurred. This is not always possible since it depends on the nature and direction of subsidence waves relative to outfall points so that provision may be necessary for as many outfall points as possible. Licences may have to be obtained to increase the capacity of watercourses outside the boundaries of the works in order to achieve this. If there is difficulty in providing a satisfactory final gradient, emphasis must be placed on watertightness and flexibility in order to maintain an effective conduit. French drains comprising perforated clay pipes with plastic-sleeved joints to ensure continuing efficiency of the pipeline during subsidence have proved satisfactory. As these pipes can be obtained in sizes of the order of 300 mm with holes provided on half of the circumference, they can be used as combined french and carrier drains by laying them with the perforated half uppermost. Shallow manholes may be constructed, using precast concrete rings provided with spigot and socket rubber ring joints which allow limited flexibility. Manholes of depths greater than 5 m should be either of monolithic reinforced concrete or precast concrete bolted segments with all joints caulked for watertightness. Culverts should be constructed, using pipes with mechanical joints or they may consist of heavily galvanized and corrugated steel pipes or pipe arches, having bitumen protected inverts. Glass reinforced plastic pipes may be used in areas subject to severe differential settlement. Careful consideration may be given to increasing the size of these culverts so as to permit lining them in the future.

Table 2

	Tension	Compression
Drystone base	$1 \cdot 0 \times 10^{-3}$	$1 \cdot 7 \times 10^{-3}$
Coated macadam base	$1 \cdot 2 \times 10^{-3}$	0
Asphalt surfacing	$1 \cdot 2 \times 10^{-3}$	$2 \cdot 8 \times 10^{-3}$
Coated macadam surfacings	$1 \cdot 2 \times 10^{-3}$	$3 \cdot 8 \times 10^{-3}$

Kerbing, channelling, footways and verges

Where possible, kerbs and channels should be omitted from new works until final settlement has occurred. If, however, they are essential, they should be of precast manufacture, with wide joints left between the units matching the gap in the bedding and backing concrete, but if severe subsidence is envisaged, it is advisable to provide 25 mm gaps every 3 m, and to fill these gaps with a waterproof compressible material.

In situ asphalt kerbing laid on the road surface by an automatic kerbing machine will take up the shape of the road as it settles without buckling or cracking. No special measures need be taken for footways of flexible construction and verges. Concrete paths constructed in in situ concrete, with frequent joints, give little trouble, but flagged paths which can be dangerous to pedestrians and are expensive to maintain in subsidence areas should be avoided if possible. Flexible paths of bitumen macadam are probably the most economical from the maintenance point of view. Although cycle tracks are rarely provided on new roads now, there may be cases where they are considered necessary, and flexible construction should be used.

Road surfacing

The use of a rigid form of road pavement is not recommended for high speed roads in areas of known or anticipated mining subsidence. The carriageway pavement should preferably be of the fully flexible type with bituminous bound base and surfacing, or if economy dictates otherwise, the base may be constructed, using dry or waterbound granular material. The type of surfacing may be either a macadam or asphalt material but where subsidence is likely to be excessive or take place shortly after opening the road to traffic, a temporary bitumen macadam surfacing may be the most economical proposition delaying the laying of the final surfacing until subsidence has ceased. Where it is likely to be minor in nature, or is not expected to take place for some time, and the road is a high speed major route, it is preferable to lay the final surfacing of asphalt and accept the rather more costly remedial work which may eventually be required.

Experience has demonstrated that various materials will withstand the range of strains shown, with insignificant damage (Table 2).

Fencing and other ancillary works

Quickset hedging or post and wire fencing is unlikely to be seriously damaged by subsidence and is therefore preferable to timber post and rail or concrete fencing of any kind. All signposts should be sited well clear of the carriageway and if possible, made of steel or timber and not concrete. Gantry-type signs spanning the carriageway should be provided only where absolutely necessary, in which case, constant surveillance is essential as they are a potential hazard to traffic if subsidence is likely to be severe. Facilities for rapid removal and re-erection must be incorporated in their design. Public utility services should be designed so as to reduce to a minimum interference with the highway when making good subsidence damage to the apparatus.

Maintenance of highways during periods of subsidence

Highway authorities in mining areas within the United Kingdom have had long experience in this type of maintenance work and an engineer faced with this problem in developing countries is advised to seek advice from the staff of these authorities.

Inspection and keeping records

Highway supervisory staff maintain a continuous watch in mining areas as do the police. It is probable that visual inspection at four-monthly intervals would suffice to reveal incipient defects with annual inspection for monitoring settlement for records. For recording the amounts of settlement, permanent ground markers embedded in the road surface are required and record plans showing their position and road profile together with any other noted defects maintained. The riding quality of the carriageway may easily be measured by the use of a profilometer. Minimum safe alignments must be maintained at all times unless speed restrictions are agreed with the police when signs must be erected to warn traffic. Where active mining operations are in progress, it is essential to maintain appropriate signs as are described in the Government Traffic Signs Manual. However, in severe cases, they should be backed up with specific signs, illuminated where thought advisable and they must be agreed with the police.

RAILWAYS

The general principles of ground subsidence are similar in all respects to those encountered in the construction and maintenance of the high-speed modern motorways and the same predictions as those described in Chapter 2 are applicable.

Early and accurate predictions of the effects of mineral extraction are of the greatest importance, since railway timetables are prepared well in advance of any notice being received of the proposed workings from the National Coal Board or other parties. Subsidence can affect the running track as follows: types of traffic involved, speeds, the type and construction of the track, the restrictions which may have to be imposed on traffic, and preventive and remedial works—such works, under the headings of preventive and remedial measures, cannot be put in hand at short notice.

In general, railways are not so interested in the total amount of subsidence, as in the relative subsidence along a length of track or within a structure. Considerations, therefore, have to be given to whether the workings are at shallow to medium depths, and these can include coal, ironstone, fireclay, gypsum and salt, but in the deeper seams, coal is usually the mineral to be extracted. All coal mining by the National Coal Board is covered so far as British Railways are concerned by Statutory Provisions, and there exists a National Agreement in this respect. There is always a full exchange of information, the basic approach being the same in practical and technical senses, whatever the mineral, but the relative legal standing of the parties will vary.

General approach

Nature of extraction. In shallow to medium depth workings the minerals can be coal, ironstone, fireclay, gypsum, salt etc. but in the deeper seams, coal is usually the mineral to be extracted. All coal mining by the National Coal Board is covered, so far as British railways are concerned, by statutory provisions, a national agreement between the two parties and a full exchange of information. The basic approach is the same, in the practical and technical senses, whatever the mineral, but the relative legal standing of the parties will vary.

The railway mining engineer's role. When a notice to work is received, the mining engineer examines the mine owners' proposals while, at the same time, the civil engineer assesses the existing situation within the railway's 'Area of protection', including establishing pre-subsidence levels and conditions. The mining engineer then forecasts, as accurately as possible the surface effects which will follow the proposed extraction. The anticipated rate of subsidence is a vital factor in predicting the probable effects on traffic operation and the extent of remedial works. The mining engineer uses a system of classification which ranges from 'very gradual' (within a range 0·50 mm per month) to 'very rapid' (exceeding 400 mm per month).

The railway civil engineer's role. It is necessary for the civil engineer, after consulting railway operations and other interests, to apply his judgement to alternative remedial measures related to varying restrictions placed on traffic so as to arrive at the primary decision; whether support should be reserved at cost to the railway undertaking and with loss of valuable mineral for the mine owner. Only in cases of exceptional difficulty is support

reserved and, when it is not reserved, then the secondary decision as to the
quality of train service which it will be possible or economical to maintain
must be taken. The cost of any preventive and remedial works must be
assessed and their implementation planned and programmed well in advance of the extraction.

Effects on railway operation of depth of extraction. Subsidence effects on track
are immediately reflected in the quality of a train service, by a reduction of
maximum permissible speeds or, in extreme cases, by requiring a temporary
suspension of the service. Shallow seams, with long-wall type of extraction
create the most difficult conditions in that they produce rapid and severe
differential settlements. The worst possible circumstances occur when a
local collapse in shallow workings is transmitted directly to the surface,
creating the so-called 'plump'. This phenomenon, usually associated with
old, uncharted workings, is the most difficult of subsidence effects to predict,
but if it can be predicted by actual location of voids below ground it is
probably the most straightforward to deal with in engineering, if not finan-
cial, terms. Multiple seams at shallow depths cause considerable damage,
not only to the track itself and to structure, but to embankments and
drainage. As depth increases, so time-scale and surface effects improve from
the track point of view. These improvements, coupled with modern mining
methods, make prediction and planning easier. Nevertheless, even with deep
seams and modern techniques, subsidence effects are not predictable to the
standards of accuracy required by a high-speed railway and small changes
in surface levels must be monitored and extrapolated from the outset.

Effects on track. Jointed track can accommodate itself fairly readily to
ground subsidence provided regular attention is given to local changes in
level ('top'), local misalignment and adjustment of joints. Continuously
welded track must, however, be broken up by 'expansion switches' or temporarily replaced by jointed track before subsidence begins. In both cases
sufficient work must be done to maintain an acceptable arrangement of
planes and vertical curves, the acceptability of the latter depending on the
type of traffic and its speeds. It is unusual for track levels to be restored, in
full, to their original values; clearance to overbridges, limitations of ballast
depths on underbridges, and the consequence of overballasting on embankments all mitigate against this. Nevertheless, in cases where the best at-
tainable profile, after subsidence, would require permanent speed restrictions
unless bridges were reconstructed and embankments widened, then these
works must be undertaken. It is important to appreciate that continuous
'band' structures such as railway track and its associated drainage suffer
reversals of stress and strain during subsidence. The initial convex strains

and tensile stresses set up during the period of 'draw' are followed by a neutral stress period and then a phase in which concave strains and compressive stresses are present. This is the most potentially damaging phase so far as the track is concerned. Whilst ballast stability has to be disturbed to permit local levelling and realignment, this same stability is essential to resistance against buckling. On lines electrified by the overhead system, local restoration of 'top' must be accompanied by adjustment of contact wire height, wherever the latter is near the minimum permissible. This, too, may lead to the raising or reconstruction of an overbridge.

Effects on structures. The data provided by the mining engineer are also used by the civil engineer to forecast the likely effect of subsidence on each structure within the notice length. Three questions need to be answered:

(a) Will the structures suffer any detectable damage?
(b) Is it necessary to carry out work in advance of mining so as to limit the probable damage or preserve the stability of the structure? This type of work is referred to as 'preventive'.
(c) Will the final condition require permanent 'remedial' works which may involve complete reconstruction?

Temporary or permanent centring introduced into masonry arches and jacking pockets in the abutments of deck bridges are forms of 'preventive' works. So, also, is the cutting of movement joints in long masonry structures or elements of structure. It is not unknown for the parapets of long masonry viaducts to 'explode' during the compressive phase of a subsidence and it is now standard practice to cut such parapets into suitable independent lengths before mining occurs.

In all cases, interim preventive measures have to take account of the fact that steps taken to counteract the initial tensile strains can be of no value in the subsequent compressive phase, e.g., the too-early grouting of a crack in masonry can lead to later crushing, whereas the crack may well close during compression. Reservation of support for a structure is only effected as a last resort but all facets must be examined. A bridge over a river, for instance, may in itself be capable of withstanding subsidence but its overall stability may be threatened if the regime of the river is severely altered.

Negotiations with the mine owner. There is often considerable scope for re-arranging the sequence of mining operations or replanning the proposed lay-out of the workings so as to reduce the effects at the surface. All the foregoing aspects having been considered, the full costs of all work anticipated must be calculated and form the basis of negotiations with the mine owner, within the framework of relevant statutes and agreements, either existing or to be drawn up.

AIRPORTS

The problems met in the construction of new runways in areas of either natural or man-made cavities are very similar to those met in the construction of modern high-speed roads and the techniques mentioned previously are applicable. In the case of concrete runways the bridging effect of the concerete will probably be of some assistance, if only to provide forewarning of a potential failure.

The possibility of active mining being undertaken beneath an existing airport can be virtually discounted but problems may arise due to uncharted shallow mine workings. In this event they should be dealt with in accordance with the principles stated earlier in this chapter for new highway construction but having regard to the lower permissible tolerances for surface irregularities.

With the exception of the crown of a camber or across drainage channels, typical specifications for new works require the finished surface of the pavement wearing course to be of such regularity that when tested with a 3 metre straight-edge placed anywhere in any direction on the surface, there is no vertical deviation greater than 3 mm. Additionally the surface shall not depart by more than 6 mm from a 45 metre optical straight-edge. Furthemore there are strict limitations on changes of grade along a runway.

CANALS

Effects of mining on waterways

Waterways are under the jurisdiction of the British Waterways Board, responsible for the control and management of 2000 miles of inland waterways. These waterways are either artificial canals or natural rivers (with stretches improved by artificial channels and structures). Such waterways may already have suffered extensive past mining subsidence damage. An essential fact governing waterways is that a water surface approximates to a horizontal plane and cannot be graded up or down like a railway or road. The dangers from subsidence damage are water overtopping subsided banks to cause flooding with the possibility of a breach and the stopping of navigation, also damage to structures, making them difficult to operate. This means that the banks and other works which support the water level have to be raised for the amount of subsidence as, of course, have any bridges crossing the water.

Status of waterway

It is necessary to establish legal and physical obligations, responsibilities, duties, land ownership, individual characteristics, e.g. fluctuating water levels at flood times, as these all vary for individual waterways.

Site survey

A site survey and investigation should ascertain type and condition of existing bank protection, available freeboard, details of structures above and below the waterway, any previous raising work etc., position of any clay core plus typical cross-sections, available access and working space.

Measures for the subsidence length

Canal banks

Knowledge of the position and amounts of subsidence, tension and compression strains, permits these predicted movements to be taken into account in the increase in elevation of embankments, the object being to accommodate movement, not to resist it. Generally, this means incorporating precautionary measures for future subsidence. A close check on timing and development of subsidence ensures that the embankment earthworks are completed in advance of subsidence. The earthworks are done to a longitudinal inverted subsidence profile before subsidence occurs; subsequently, subsidence causes the raised bank top to return to its former height above water level. By this means, the navigation may function throughout the subsidence period.

Locks

Locks should be examined to ascertain whether they can withstand the effects of subsidence and be raised in overall height; some locks are already very deep. Lock walls may be subjected to substantial external water pressures. For surface buildings a small degree of superficial damage may be ignored but even a small amount of cracking at a lock can release water under pressure and induce leaching and piping, collapse of gates, culvert failure, heaving of floor and bowing of walls. The raising works may include fixing new top and bottom gates, raising the chamber walls, strengthening the floor plus alterations to ground paddles and culverts, such work usually requiring a total possession with a programmed stoppage.

Bridges

Where loss in headroom could impede navigation the bridge deck is raised or reconstructed before subsidence with in-built precautions for subsidence movement.

Tunnels

Tunnels are affected by loss of headroom, possible distortion and damage to the tunnel lining. Repairs can be difficult and prolonged. An alternative method for dealing with the damage could include removing the tunnel cover or lowering the pond.

Aqueducts

It is extremely difficult to maintain aqueducts during the subsidence period. Unless the aqueduct may be lowered progressively with subsidence, the process of lowering temporarily increases the water depth, imposing increased loads. Where an existing cast-iron or masonry aqueduct is already difficult to maintain, a new structure may be preferable.

Stop plank grooves

Provision and maintenance of stop plank grooves enables lengths of waterway to be isolated, with attention to any feeders and drains. The grooves may require to be extended upwards and the cills raised correspondingly.

Waterway channel

Water's edge bank protection. A stone or brick wall may be raised by capping in concrete or brickwork. Additional stability can be gained by driving a single steel sheet pile at intervals along the face of the wall, this being anchored back. Existing bank protection, which is inadequate for raising, may be left in place and a new form of protection introduced, say steel sheet piles positioned to suit site conditions near the existing wall (special care being necessary where rock outcrops near the surface as driving piles can induce leakages). Where ground conditions are not suitable for driving steel sheet pipes, a gabion wall can be constructed. For small amounts of subsidence, stone pitching or other suitable material, say, colliery shale, may be used.

Clay core. When raising works form embankments, clay cores are important as subsidence movement can easily cause leakages. Clay cores up to 1 m wide exist on some artificial waterways in mining areas. The clay core is raised by first removing the top 150 mm of clay and the new puddle clay material bonded into the existing clay. The new puddle clay is placed in layers not exceeding 150 mm thick and thoroughly puddled with watering as necessary. Where no core exists, it may be necessary to form a new one founded below channel bed level.

Backfill. The backfill is placed behind the bank protection and clay core. This must be inert material such as colliery waste which can be suitably blanketed to meet environmental requirements.

Channel infill. Following subsidence, the channel may need infilling with clay to control leakage under increased hydrostatic pressure. A fillet can be placed against the banks to provide temporary support during the subsidence period.

Alternative treatment. The alternative to physical raising may be to lower the water level permanently, or the water level control point may naturally subside and the level could be left permanently lowered. Some dredging of the unsubsided length may then be required. When navigation is on natural rivers the responsibility for bank raising may overlap with those of the Regional Water Authority. It may only require the raising of the towpath with the other bank being allowed to subside below water level without raising but identified by markers placed in the channel along its line.

TUNNELS

In the construction of tunnels, the same precautions have to be taken as in mining. The width of the tunnel determines the effect of movement which could take place in varying types of stratum.

In mining, subsidence is caused by the collapse of the superincumbent strata behind the working face and support has to be given to roadways by

Table 3. Magnitude of ground subsidence above shield-driven tunnels

Ground cover/ dia. $(Z/2a)$	Ground	Lining	Area ratio %
0·6	Stiff clay	Exp. conc.	0·2
5	Stiff clay	Exp. conc.	1·2
1·1	Fine sand	Bolt CI	2
4*	Silty clay	Bolt steel	5
2*	Silt	Bolt CI	6

Table 4. Direct causes of ground loss

Nature of ground loss	Normal limits, %, ground loss/tunnel area
At face	0·1–?
Behind cutting edge or bead	0·1–0·5
Along shield	0–1
Behind tail	0–4*
	0–2

*Below water-table in compressed air

the form of props or steel arches and steel shields. Similar precautions are applicable to the driving of tunnels and, as mentioned above, the width is critical.

Table 3 indicates some typical figures for total subsidence for a shield-driven tunnel expressed as a ratio between total ground subsidence and the area of the face of the tunnel, clearly valid only for reasonably long lengths of continuous working. Table 4 shows limits of the contributing factors.

REFERENCES

Malkin, A. B. & Ward, J. C. (1972). Subsidence problems in route design and construction. *Q. Jnl Engng Geol.* **5**, No. 1/2.

West, G. & Dumbleton, N. J. (1972). Some observations of swallow holes and mines in the chalk. *Q. Jnl Engng Geol.* **5**, No. 1/2.

Chapter 6
Land drainage

GENERAL

The term 'land drainage' has a wide interpretation and describes natural systems for the discharge of water originating from rainfall percolating from or flowing off land, ranging from field ditches to large rivers. It is the work of drainage engineers to control soil water to provide proper conditions for agricultural production and to improve natural water courses for the alleviation of flooding by:

(a) widening;
(b) deepening;
(c) re-grading or embanking;

and to provide protection known as defences against tidal flooding.

General application

The land drainage systems on a moderately flat area of agricultural land cause many problems due to the effects of mining operations of whatever character they may be, and these may be summarized as follows. The extraction of a seam of coal will result in the lowering of an area of the surface in relation to Ordnance datum and water levels in the surrounding water courses. The effects of the advancing waves of subsidence might reduce or reverse the gradient of drainage channels, and reduce the freeboard of embankments. These affect major watercourses with tributaries conveying the drainage from the adjacent operations, large drains which convey water from the internal catchment areas and field ditches into which tile drainage systems discharge. The watercourses and large drains may be embanked to prevent the overflow of flood water and land drainage pumps may be installed to lift water from the internal drains into carrier drains or rivers when high flood level precludes gravity discharge. The loss of efficiency in a drainage system of an area in which mining subsidence has taken place affects natural field drainage, tile drainage, reduces drainage standards, and can give rise to intermittent and permanent flooding of agricultural land and urban properties. The Land Drainage Acts of 1930 and 1961 gave a Regional Water Authority general responsibility for the supervision of land drainage, but such authority can only exercise permissive powers to undertake work to major watercourses designated as main rivers. The selection of watercourses is the prerogative of the Water Authority, and designation, after consultative procedures have been taken, is finally made by the responsible Government Department. The Water Authority may, at any time, apply for a variation to the order and the Water Authority is, therefore, responsible for the maintenance of those main rivers within that area.

Internal Drainage Districts are administered by a Board with autonomous permissive powers to maintain and improve the local drainage systems, and by a simple resolution, the Drainage Board may adopt and work any watercourse in the district. Outside Drainage Districts there are many small watercourses where maintenance rests with the riparian owner or occupier, but a Local Authority can improve such watercourses under the provisions of the 1930 and 1961 Land Drainage Acts. The effects of mining subsidence on land drainage was first considered in 1923 by a Royal Commission which presented its report in June 1927. As coal mining in the Yorkshire Coalfield progressed eastwards, problems arose within the low lying areas of Doncaster and, as a result, a special inquiry which was held resulted in the Doncaster Area Drainage Act 1929, which placed the responsibility on the mine owner to keep open the existing drainage works and do such things as would obviate flooding within the district. A similar protection was extended to land drainage interests in the remainder of England and Wales under the provisions of Section 5 of the Coal Mining (Subsidence) Act 1957, whereby the National Coal Board is responsible for remedying, mitigating or preventing deterioration in the land drainage system by reason of subsidence which has occurred, or appears likely to occur within the district. The Coal Board may elect and pay the Drainage Board for the carrying out of such works, but in the case of main rivers, they are obliged to make payment for the cost of the remedial works to the Water Authority. Remedial works may be merged with improvement works and the ultimate cost proportioned between the Water Authority and the National Coal Board.

For the development of the Selby Coalfield, the Yorkshire Water Authority has an additional agreement with the National Coal Board in extending and improving the provisions of the Coal Mining (Subsidence) Act, 1957.

RIVER DRAINAGE

The effect of subsidence on main rivers depends, to a large extent, on the physical geography of the affected area.[1] In comparatively hilly districts where a river flows in a defined valley and the bed gradient is fairly steep, subsidence may have little effect. The relative level of the river and land generally can be restored by channel improvement and regrading. It is advisable to undertake a soil investigation and calculations for the embankment stability before commencing extensive channel regradings, particularly where excavations are to take place in clays and silts.[2] In urban districts, room for widening and deepening may not be available, and support for the banks may be required in the form of retaining walls of stone filled gabions, brick, mass concrete or diaphragm concrete, timber, or sheet piling. It may be necessary to underpin existing retaining walls or bridge abutments, and this work may take the form of a reinforced concrete trough inside the abutments or mass concrete foundations, possibly in conjunction with steel piling supports and provision of tied or raking anchorages or other underpinning techniques and wall strengthening.[3] Where regrading is not practicable, or will not entirely offset the effects of subsidence, the river may be embanked or existing embankments raised prior to subsidence taking place, by calculating the amount of vertical lowering and its subsequent effects. Earthen embankments can be constructed or raised with spoil from channel improvements, borrow pits or imported colliery shale.[4] Soil tests and stability analysis will determine the type and construction of the embankment. Retaining walls or riverside structures such as jetties, may be raised and strengthened providing the resultant structure is designed to withstand the new conditions and the strains and stress induced by the effects of the subsidence. Where it is necessary to carry out structural remedial works, these should be deferred until subsidence is complete.[5] Where navigation takes place on the larger rivers, the lowering of a weir may have an adverse effect on the reduction in the depth of water in the upstream reaches which have not been affected by subsidence. The loss of hydraulic support and increase in velocity may lead to slipping on river embankment or cause channel erosion. The weir may, therefore, be required to be raised by the necessary amount of vertical lowering which takes place, but the following factors will require to be taken into consideration.

(*a*) The stability of the existing and raised structure;
(*b*) The effect of downstream erosion;
(*c*) The effect of upstream flood level;
(*d*) The necessity for progressive raising during the subsidence period;

(e) The possibility of temporary raising with permanent works to follow on completion of the subsidence;

(f) The probability of further coal seams being extracted in the future.

Where an outfall is permanently submerged due to subsidence and gravity drainage impaired by loss of hydraulic gradient or by siltation, it is necessary to rebuild the structure preferably following the subsidence.

Two problems may arise which will require the regrading of the channel, the underpinning and protection of the bridge foundations, the raising of the deck by jacking and increasing the height of the abutments and piers to restore the headroom which will allow navigation to continue. A river channel, when improved and embanked, has a finite capacity which may, on rare occasions, be exceeded. To prevent the flow of floodwater, therefore a subsided basin possibly remote from the river may require inland or retired banks to be constructed. Although not generally economical, local depressions may be raised by depositing and spreading colliery or other waste after the removal and subsequent restoration of the topsoil. Areas of low-lying subsided land adjacent to silt-laden tidal rivers may be embanked and slowly raised with silt which has settled following the controlled inflow during spring tidal cycles or, more quickly, by depositing pipe-conveyed silt dredged from the river channel. It may be expedient to improve a river channel or flood defence in conjunction with remedial work, in anticipation of, or following mining subsidence, with appropriate and generally beneficial apportionment of costs.

INTERMEDIATE DRAINAGE

Intermediate drainage channels, within internal drainage districts often have little fall, and by reducing or reversing gradients, mining subsidence generally demonstrates an immediate adverse effect on the local drainage system. The principles and methods used for river drainage can be adapted for intermediate drainage systems, but where land cannot be drained effectively by gravity or extensive regrading, is neither practical nor economical. It is necessary, therefore, to construct a pumping station to lift water from the subsided area and discharge it at a higher level to a free-flowing watercourse.

A long-term appraisal of the coalfield, particularly with multi-seam extraction is essential to determine the most economic and effective site for a pumping station. Ideally, it is preferable for mine workings to commence at the downstream end of a drain where a station might be located and progress the workings upstream, thereby creating a positive gradient. It is advisable to make a soil survey and check the stability of the ground before commencing extensive channel regrading, particularly where excavation is in clay, silt or a strata of sand or gravel. Alternatively, temporary pumps near the perimeter of the advancing subsidence trough can progress with mine workings, lifting water from subsided drains into the lengths of drain which have not yet been subsided.

A principal arterial watercourse may require a large pumping station to discharge high flood flows, and it may be more economical to construct embankments to maintain the gravity flow of upland water with pumping limited to drainage from the subsided basin.

Where multi-seam working leads to severe differential variations in the relationship of the land to the drain level, booster stations are required. The capacity of the pumping station will be determined from the size and characteristics of the catchment area, including geography, geology, development and floodwater storage, which includes the study of geography, geology and the development of the area. In order to permit the economical discharge of a wide range of flows, and to allow for maintenance and repairs, it is essential that there should be one or more pumps which can operate automatically.

The construction of a pumping station as a result of mining operations often affords the opportunity to effect an overall improvement in the internal drainage system and the benefits of merged works can be computed and apportioned between the drainage authority and the Coal Board. It is essential that both the river authority and the Coal Board work in complete liaison with each other, in order that the site of the new pumping stations may be so located to take care of future lowering due to further mining operations.

FIELD DRAINAGE

Experience has shown that if the arterial drain or ditch systems are dealt with, as described in the foregoing section, large areas of agricultural land can be undermined without adverse effect on the field drainage system.

Natural drainage

Water can drain naturally from land by following the slope of the ground until it discharges into an intercepted watercourse. It can also percolate through the ground surface or downwards through fissures to the elevation of the seasonal level of which the water table is governed by geological and soil characteristics and the depth of the nearest watercourse. Under some circumstances, differential settlement will cause the water table to approach the land surface at the position of maximum lowering, and in extreme cases, water may rise above the surface and create ponding.

Pipe systems

Pipe drains are installed to control the water table and remove surplus water from land which is lacking sufficient gradient, or soil which is not sufficiently permeable to be free-draining. Modern pipe drainage systems are constructed from porous fireclay, concrete pipes or perforated plastic piping. The pipes are usually surrounded by free-draining granular fill material. The depths of the pipes are related to the elevations to which the groundwater table must be lowered and to the nature of the crops to be grown. The spacing of the pipe drains is related to the depth of the pipes and the permeability of the soil. A minimum recommended gradient is 1 in 700 but gradients less than this may be encountered.

Pipe drains are not generally susceptible to damage because of the nature of their construction. Butt-jointed pipes or plastic piping can accommodate mining movements without experiencing physical damage or significant changes to the horizontal or vertical alignment of the pipes. An exception to this general case could arise where a drain is located at or near the surface position of a geological fault, or where mining is carried out on a partial extraction basis, leaving small pillars between the extracted areas. Differential settlement of the surface can cause the pipe drains to sag and retain the water. The physical effects of this on the surface will be ponding, unless the drains are re-aligned in relation to the subsidence caused by the mining operations.

It is difficult to prevent adverse effects on existing pipe drainage systems, particularly where the receiving watercourses have not subsided the same amount as the pipe drainage system. It means that extensive new draining has to be carried out and in some cases the Coal Board will have to pay a depreciation payment for the land which is in an area where no remedial measure can be untertaken, resulting in permanent ponding.

Mole drainage systems

A mole drainage system will be affected in the same way as other pipe systems and the same types of remedial work should apply. The mole drains themselves will be subject to the effects of tilt which will alter their gradients. They are, however, divided into short runs by frequent intersections with main drains, and so they are likely to be able to continue to function so long as the main drain stays empty. If there has been any long-term waterlogging of the mole drain, it is likely that they will have collapsed. Under normal circumstances, new mole drains have to be drawn every five to ten years and, if there is any doubt as to their condition after subsidence has taken place, a new mole drainage system should then be installed.

REFERENCES

British Standards Institution (1956). BS2760. *Pitch-impregnated fibre drain and sewer pipes.* Brit. Stand. Instn.

Cole, J. A. (1960). The detection of leaks in water mains by nitrous oxide. *Wat. and Wat. Engng* **64,** 453–454.

Cole, J. A. (1961). An economic analysis of systematic leak detection. *Jnl Instn Wat. Engrs* **15,** 429–444.

Farran, C. E. (1952). The effect of mining subsidence on land drainage. *Jnl Instn Wat. Engrs* **6,** 482.

de Lathouder, A. (1961). Derivation and consideration of some equations describing the problem of leak detection. *Jnl Instn Wat. Engrs* **15,** 445–459.

Orchard, R. J. (1957). The effect of mining subsidence upon public health engineering works. *Jnl Instn Publ. Health Engrs* **56,** 188.

Chapter 7
Services

INTRODUCTION
General
Those services which are dependent on underground systems present special problems in respect of ground movement. The pipes and cables of such installations can be affected more severely than many free-standing structures because of the continuous nature of the former and their anchorage within the ground. This is particularly the case where subsidence results from the active extraction of minerals by long wall mining methods and the buried services are subjected to the entire ground movement components. In such circumstances, the main effects on services which have to be considered are the temporary and permanent compressions and tensions, the alterations to gradients (mainly for gravity drainage systems) and, where large-diameter pipelines of considerable length are concerned, curvature and displacement.

Other forms of ground subsidence will tend to be largely unpredictable as for example that arising from the localized collapse of old shallow mine workings or the widespread effects of water or brine extraction. These situations normally preclude or make difficult the incorporation in service systems of any realistic precautionary measures to accommodate the effects.

It is obviously preferable to consider the question of precautions at the design stage although it is possible to incorporate certain measures into an existing system when the possibility of damage is anticipated. Precautionary measures should be related to the worst possible situation that can arise in the given circumstances unless there is reliable evidence to the contrary. This is particularly relevant to present-day mining where, because of the generally large extent of extraction, the mining operations that will induce the movements are likely to be of a long-term nature. Broadly, the aim of designing precautionary measures should be to ensure that in addition to reducing damage and the consequent cost of repair work, every effort is made to avoid any disruption of the service itself. In districts where ground movement is occurring or is likely to occur, comprehensive and regular inspection surveys should be made where possible.

The effects of ground movement on pipes can be summarized as follows:

(a) flexural movements tending to rupture the pipe transversely;
(b) axial shortening tending to rupture the pipe longitudinally or to cause buckling;
(c) axial lengthening tending to cause drawn joints (especially at bends, branches, manholes or valves); these effects can be aggravated by local resistance to sliding due to projections beyond the pipe barrel or other anchorages;
(d) Severe transverse shear forces.

Orientation

Subsidence effects on pipes and cable will also vary in relation to the direction of extraction. A pipe which lies transverse to a working long-wall panel will be subjected to definable zones where compressive and tensional strains will build up. This condition will remain after the working has stopped or passed out of the area of influence. On the other hand, a pipe lying parallel to the line of working will be affected by a travelling wave of movement which will induce temporary stress and displacement but which will eventually leave the pipe in a relatively strain-free condition. Variations in this type of situation will of course occur where the line of the installation itself changes direction; or when a series of long-wall panels is being worked; or where there is a significant change in the character and strength of the intervening strata. In addition the possible complication of multi-seam extraction may require to be taken into account.

Faulting

A special problem can arise in mining situations when the normal development of movement is affected by the presence of large geological faulting. This may result in severe localized effects being set up at the position of the fault outcrop and this usually takes the form of a step or abrupt change in the ground profile. The large flexural and shearing forces that are likely to be induced in such circumstances can cause severe damage to service installations. It is also possible, although unusual, to have circumstances where bedrock is locally close to the surface thus giving rise to similar localized severe movement.

Pipeline behaviour

The manner in which a pipeline responds to the various movements and forces to which it is subjected will depend on:

(a) the strength, thickness and flexibility of the pipe material;
(b) the spacing of flexible joints where these are used and their ability to accommodate displacement of the pipe in any direction;
(c) the degree of axial anchorage, if any, imposed on the pipeline;
(d) the depth of cover and type of material in which the pipe is bedded;
(e) the age and condition of any existing pipe.

Choosing the most suitable material for pipelines must be a compromise between strength and ductility, resistance to corrosion, and cost relative to the design life of the pipe and local conditions.

Pipeline requirements

To meet the special conditions attached to choosing a pipe for a ground movement situation, certain preferences can be recognized. Basically, the aim should be to provide a pipeline which has the requisite degree of flexibility either in material properties or in an arrangement to suit the anticipated ground movements. Suitable joints should be used which will accommodate the necessary angular and telescopic movement.

Welded steel pipes are relatively flexible, have generally considerable resistance to buckling, and are usually well able to tolerate ground subsidence. Their smooth internal surface obviates the flow problems that can arise with joint grooves. The absence of outside projections greatly reduces the frictional resistance of the trench material experienced by conventionally jointed pipes. It is essential, however, to ensure that if a welded pipe system is chosen, the joints are rigorously tested on installation.

Polyethylene pipes can bend freely and suit many situations of ground movement. Pitch fibre and unplasticized PVC pipes have good load–deformation characteristics and while they may perhaps have a limited capability for axial extension they can now be fitted with suitable telescopic joints.

Pipe joints

There are several special pipe joints on the market and it is necessary here only to point out some basic factors relating to their installation. Care should be taken to ensure uniform bedding or anchorage of a pipe run in order to obtain, as far as possible, even distribution of strain throughout the joints. Alternatively, a coupling harness can be used in addition to the sliding joint which prevents complete pulling out of the pipe sections.

Pipes fitted with flexible joints should not be bedded in concrete which may break in random fashion causing local concentration of strain. It is preferable to use a granular material for bedding such as broken stone or gravel. If a concrete base is unavoidable owing to special requirements then suitable transverse breaks should be made in the concrete at every flexible joint.

The number of flexible joints which should be provided in any pipe run should be such that their total possible movement is equal to the predicted change in ground length together with due allowance for thermal expansion.

It is essential to ensure that in laying the pipeline, none of the available capacity for flexibility is lost by poor assembly. To achieve this may involve a higher standard of supervision during installation with the inspection of joint spacings by means of templates, gauges or the use of suitably marked spigots and pipe ends. Where steel pipes have been fitted with a cathodic protection system to reduce corrosion, it will be necessary to ensure that electrical continuity is maintained if flexible joints are installed.

The performance requirements specified in the appropriate British Standard documents on pipe joints are sufficient for the types of movement likely to be experienced in all but the most extreme of mining situations. In cases where very severe subsidence is anticipated it may be necessary to shorten the individual pipe lengths.

A final general point in respect of pipe laying in areas of ground movement is that flexible joints should be fitted in pairs at all locations where service pipes enter or leave manholes or structures.

WATER SUPPLY
General
The basic effects of ground movement on the pipes of a water supply system are broadly similar to those listed in the preceding section. In addition it has been noted that trouble can occur where the gradient between scour and air valves has been reduced or reversed. A further common source of trouble with the smaller water supply systems is at the points where the ferrules for service pipes have been tapped into the mains. Apart from the factors discussed previously for pipelines in general, the presence of branches, junctions and valves introduces a further element of possible vulnerability. The age and condition of water pipes, particularly where high pressures are involved can be important contributory factors in terms of possible damage.

Water pipe precautions
Large water mains in areas likely to be subject to movement are now normally laid in steel pipes with Johnson-type couplings. Where smaller mains of up to approximately 450 mm diameter are involved spun iron pipes with mechanical joints can be used.

Where a water main consists of old cast iron pipes with run lead joints it can be expected to react badly to subsidence. In such a case, and particularly if the pipe is known to be in poor condition, it may be advisable to take precautionary measures prior to movement taking place. Either flexible joints could be installed or in extreme cases the laying of a temporary link on the surface to bypass the length of pipeline likely to suffer damage.

The protection of steel pipes against corrosion presents an additional problem in that movement of couplings can expose bare metal while the bending of the pipe itself could crack any brittle coating. Zinc treatment during manufacture is one method of catering for this difficulty.

In areas of subsidence, it is advisable to lay water mains at as shallow a depth as possible and to bed the pipes in soft material. Apart from helping to reduce the amount of transferred movement, this also facilitates access for inspection and repair. A precautionary measure adopted by some authorities is the fitting of self closing valves on all mains of large diameter. This practice is not advised however on pumping mains. Where a water main is tapped by service pipes, the joint should be made in lead with a loop at the ferrule to permit movement. The regular inspection of all water pipes which are liable to be affected by ground movement is a further desirable precaution.

Aqueducts
The location of aqueducts and the alternative sources available, should their water supply capabilities be affected, will influence the question of precautionary measures. In some cases, where the water is being conveyed through a tunnel or over a bridge, the only practical answer may be to purchase mineral support. If the aqueduct consists of a number of parallel pipes then the provision of cross-connections may allow the supply to be maintained should any of the pipelines suffer damage through subsidence.

Other precautionary measures additional to those recommended previously for water mains which may be desirable in the case of aqueducts are the provision of large air valves to facilitate charging up after repairs, and alarm bells operated by self acting valves, water levels or pressures.

Boreholes
Working minerals under or in the vicinity of a water supply borehole can cause a distortion of the hole itself which can affect the pump (if submersible), the supply pipe or the borehole lining. This possibility can be minimized by carrying out the extraction of the mineral in such a fashion that differential vertical lowering will be kept to a minimum. Again, con-

sultation with the mineral undertaker at an early stage is necessary, if action of this type is contemplated, as mining considerations may make such a scheme impractical or too costly. If this should prove to be the case then it may become necessary to negotiate for some form of support in order to keep the amount of movement within acceptable limits. With new boreholes, consideration should be given to the desirability of extending the lining further down the hole than might ordinarily be considered necessary. This would help prevent the possibility of jamming by rock which may be dislodged by ground movement. As an ancillary point in connection with water supply boreholes, it is worth noting that, if the ground movement arising from the extraction of minerals in the vicinity is considered likely to be severe enough to cause serious distortion, then it is probable that the water source itself will be affected. This could either take the form of a loss or reduction in the supply because of a general lowering of the water table, or impermeable strata may be fractured so permitting the inflow of water containing undesirable salts or other forms of contamination.

SEWERAGE
General
The general alteration in level induced by ground movement is by itself unlikely to be critical in respect of most forms of service installation. It is of fundamental importance, however, when considering sewerage systems. In addition to the general problems already referred to of cracked and broken pipes and pulled joints, subsidence can result in sewer gradients being flattened whereby pipes lose their designed capacity and become prone to silting. Reversal of gradient can form inverted siphons, blockage and flooding from manholes. The original design characteristics of a sewerage system can by these means become seriously affected.

A further special problem, particularly where main trunk sewers are involved, is the risk of pollution arising from cracked pipes. Depending on the extent of fracturing and the soils condition, this can result in either a slow seepage and the infiltration of a wide area of ground which may not be apparent for some time or it can rapidly cause gross pollution of land and watercourses. Conversely, cracked pipes may allow the entry of groundwater into the sewerage system and while there will be few cases where the volume of incoming water will seriously affect the pipe capacity, there is a risk that the removal of water from the land or from superficial deposits will give rise in itself to surface settlement problems. Inflow may also remove fines from the ground and cause local cavitation leading to possible sudden surface settlement.

Precautionary measures

Ideally, sewers in areas subject to ground movement should be designed with sufficient capacity and gradient to overcome any adverse effects, if and when they arise. In practice it is rarely possible to forecast all the requirements of a system and attain this policy of perfection. Indeed it seldom happens that site conditions and cost factors permit such generous margins to be incorporated as to eliminate all risk of adverse gradient.

Nevertheless, early attention to details of possible settlement may enable a measure of compromise to be obtained and so reduce the size of any eventual problem. Every attempt should be made at the outset to avoid the situation where the ultimate solution can only be the expensive one of permanent pumping. It may be that the extra initial cost of opting for a different route for the laying of a main sewer so as to avoid an area of settlement, will prove to be more economic in the long term.

If possible, the separate system of sewerage should be adopted. The foul sewer will obviously be smaller for the separate system, thus tending to facilitate repair work or relaying if this should become necessary. Moreover, entry of grit into the foul sewer is minimized and this factor, together with the increased probability that self cleansing velocities could be maintained, would in turn minimize the possibility of blockage if the original gradients were affected adversely.

It would appear preferable therefore to give consideration to the use of reinforced concrete, prestressed concrete or special thick-walled concrete sewer pipes in areas of ground movement. Alternatively steel pipes might be used. Where subsidence conditions are not expected to be severe and extra strength pipes are not available, consideration could be given to the use of normal fireclay or plain concrete pipes embedded in a continuous reinforced concrete surround with suitable hinges and expansion/contraction joints at junctions and at intervals along the sewer.

Sewer manholes should be of reinforced concrete or precast concrete segments surrounded by six inches of reinforced concrete, the use of brick being avoided except for the adjustment of covers or any special form of construction. Generally, manholes on small sewers are of precast reinforced concrete while for large diameter sewers in situ reinforced concrete is an excellent material. The joints between sewers and manholes should be flexible. Where the anticipated ground movement is likely to be long-term it may be possible in certain circumstances to leave a main sewer in open cut supported by a series of cradles which have facilities for jacking.

ELECTRICITY SUPPLY

General

The supply of electricity by either cable or overhead line has not proved to be particularly prone to damage by ground movement. This is probably because there is usually sufficient tolerance in both systems to cater for the strains set up by minor subsidence. There are, however, certain factors which must be recognized as relating specially to this service and particular care must be taken to ensure that in areas where severe ground movement is likely to arise that all possible precautions are taken.

The cost of repair of a broken high voltage cable may not be excessively high, but the social consequences of an abrupt power failure can be very serious indeed.

Underground cables

Effects of movement on cables. In the majority of cases, damage to buried cables takes place at joints when the ground subsidence has been severe enough to cause a movement between the internal conductors and the external sheathing. Where the cable is subjected to tensional effects, the core jointing may be completely drawn thus causing a discontinuity. Compression may cause buckling of the conductors giving rise to earthing by contact with the lead sleeve over the joint in the sheathing. In severe cases the lead sleeve itself may be broken. Where lead-sheathed cables are used, the sheath may be fractured sufficiently to permit moisture to enter the cable. With paper-insulated cables distortion may cause tearing of the paper. In both cases the damage can result in electrical failure. Certain types of cable, chiefly those of the oil-filled or gas compression types used in high voltage systems, sheath damage does not necessarily result in cable breakdown provided arrangements exist for the replenishment of the oil or gas at a rate sufficient to offset the loss at the point of damage. On high voltage cables the faults usually become apparent by the operation of protective gear which allows fault location procedures to be set in motion but on low voltage systems where ring mains are common, open circuits may not be evident for a considerable time. A further hazard with low voltage cables is the possibility of broken neutral conductors although this is normally guarded against. Underground link disconnecting boxes may develop faults; they have been broken on occasion by movement of the associated cables.

Precautionary measures with cables. A common procedure where ground movement is anticipated is to provide a measure of slack in the cable by laying it in a bed of sand and snaking it from side to side of the trench. The same principle but using a wide concrete duct is likely to prove more effective but is of course much more expensive. With high voltage cables in particular, however, it is preferable to use cables with a double layer of steel wire armouring to resist abnormal tension effects. This can be combined with the use of an expansion joint which permits horizontal movement of the conductor cores. Joints can be protected by incorporating expansion pits on each side of the joint. The cable is snaked into these pits and this gives scope for movement without damaging the joint itself. The pits can be made accessible for inspection and measurement of movement. Where expansion joints are not used on medium or low voltage cables, the neutral conductor may be provided with an over length jointing ferrule in a normal joint so that the phase conductors will break before the neutral thus avoiding unbalanced voltages on single phase services.

Overhead lines
Wooden pole lines are only likely to be affected by severe movement which will tilt or move the poles sufficiently to cause a tensioning of the conductor. This situation can usually be readily seen and the appropriate measures taken to alleviate the trouble. Steel towers normally employ suspension type insulators to carry the conductor wires. Displacement of the tower top may result in reduced clearance between the structure and its hanging conductors with the possibility of electrical flashover.

As a result of the local subsidence of one tower in a line, or the tilt component itself, a reduction in conductor ground clearance below the allowable statutory limit may result. Because steel towers are usually built on independent concrete blocks for each leg, and do not have high structural strength they are susceptible to distortion in the event of differential ground movement. It is sometimes possible initially to plan a route to avoid ground subject to severe subsidence. If serious movement is anticipated, the pylons can be built with jacking facilities incorporated in the foundation bases. The tower structure can be strengthened in its general members at the design stage or an existing tower can be fitted with collar beams round the base to prevent the legs from spreading.

As angle towers carry heavier loads, their tolerance to movement can be more limited while the change in direction of the conductor line imposes a further complication to the tilt factors. In addition, they have less scope for making adjustment to cable catenaries. Line routes which are being affected by ground movement should be carefully patrolled so that the effects can be detected and monitored and the appropriate steps taken to minimize damage.

TELECOMMUNICATIONS

The comments made in the foregoing section can also be applied to most aspects of telecommunication systems such as cables, telegraph poles, radio and television masts etc. The ducts carrying Post Office cables are normally capable of deflexions which can accommodate considerable movement before the cables themselves are affected.

Certain apparatus contained within telephone exchanges can be sensitive to changes in gradient and, where possible, provision should be made in new buildings to cater for this aspect by either structural or apparatus design. It may be possible to overcome the problem in existing exchanges by the continual resetting or jacking of internal support frames.

GAS AND OIL SUPPLY

The general comments on pipe installation made previously in this chapter, apply equally to gas and oil pipelines. Also there should be little need to emphasize the obviously serious nature of even a slight continuous leak from a cracked pipe or a pulled joint. In areas of subsidence and particularly where the gas supply system consists of old cast iron pipes there is therefore a special need to carry out regular inspections and to replace faulty or suspect joints and valves. New gas mains in areas with a high risk of movement should be laid in either steel pipes with flexible joints or in welded steel pipes. Joints between steel and cast iron pipes should be made with stepped flexible couplings. Flexible joints should also be provided at service connections to the mains and at valve positions. Main distribution pipelines for gas and oil are now invariably constructed of welded steel and as such are capable of considerable movement.

REFERENCES

McCallum, T. (1943). Mineral subsidence and local authority services. *Proc. Instn Munic. Engrs.* **70**, 441.

Serpell, C. A. (1949). The laying of a steel pipeline. *Jnl Instn Wat. Engrs.* **3**, 17.